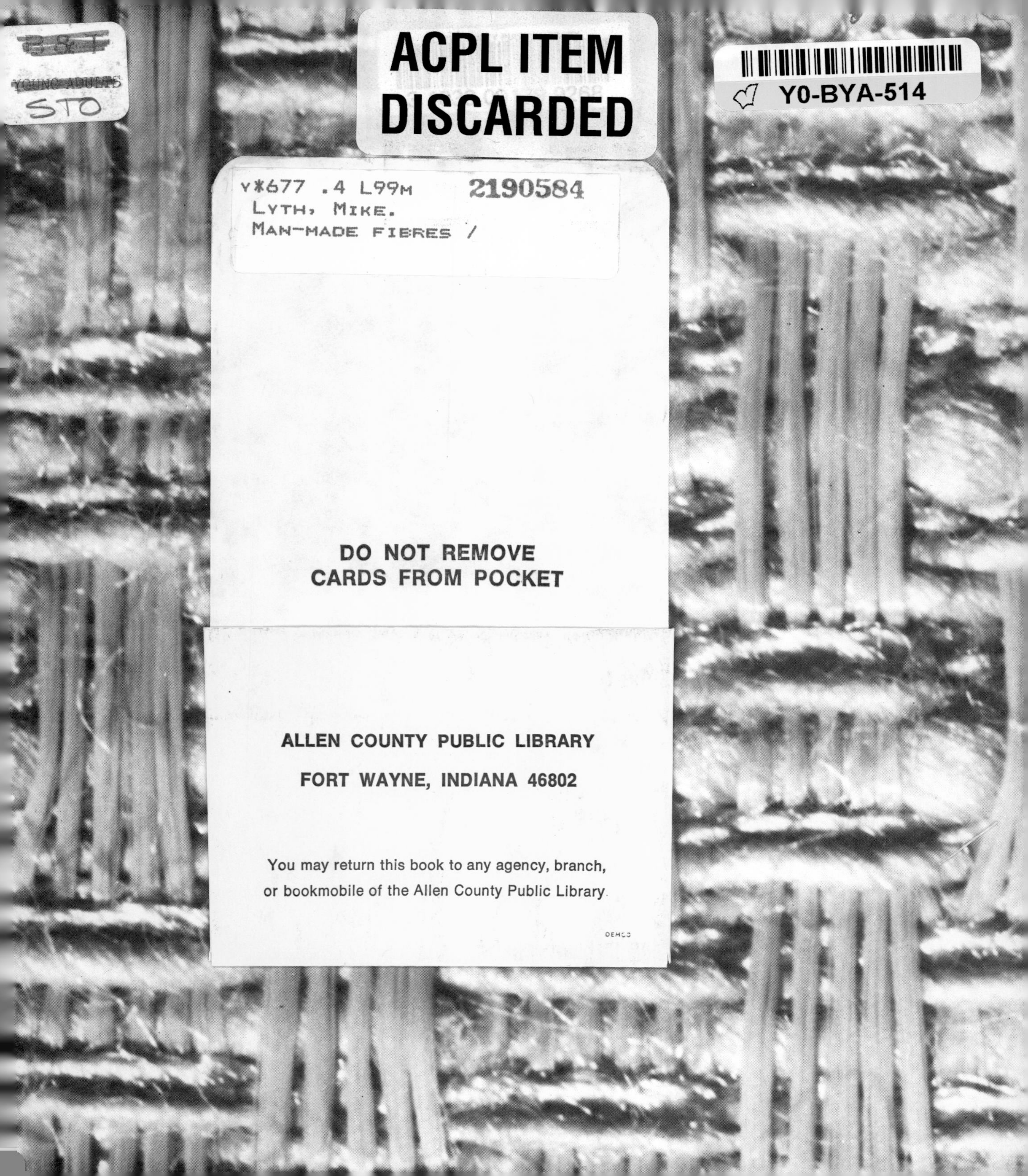

Man-made Fibres

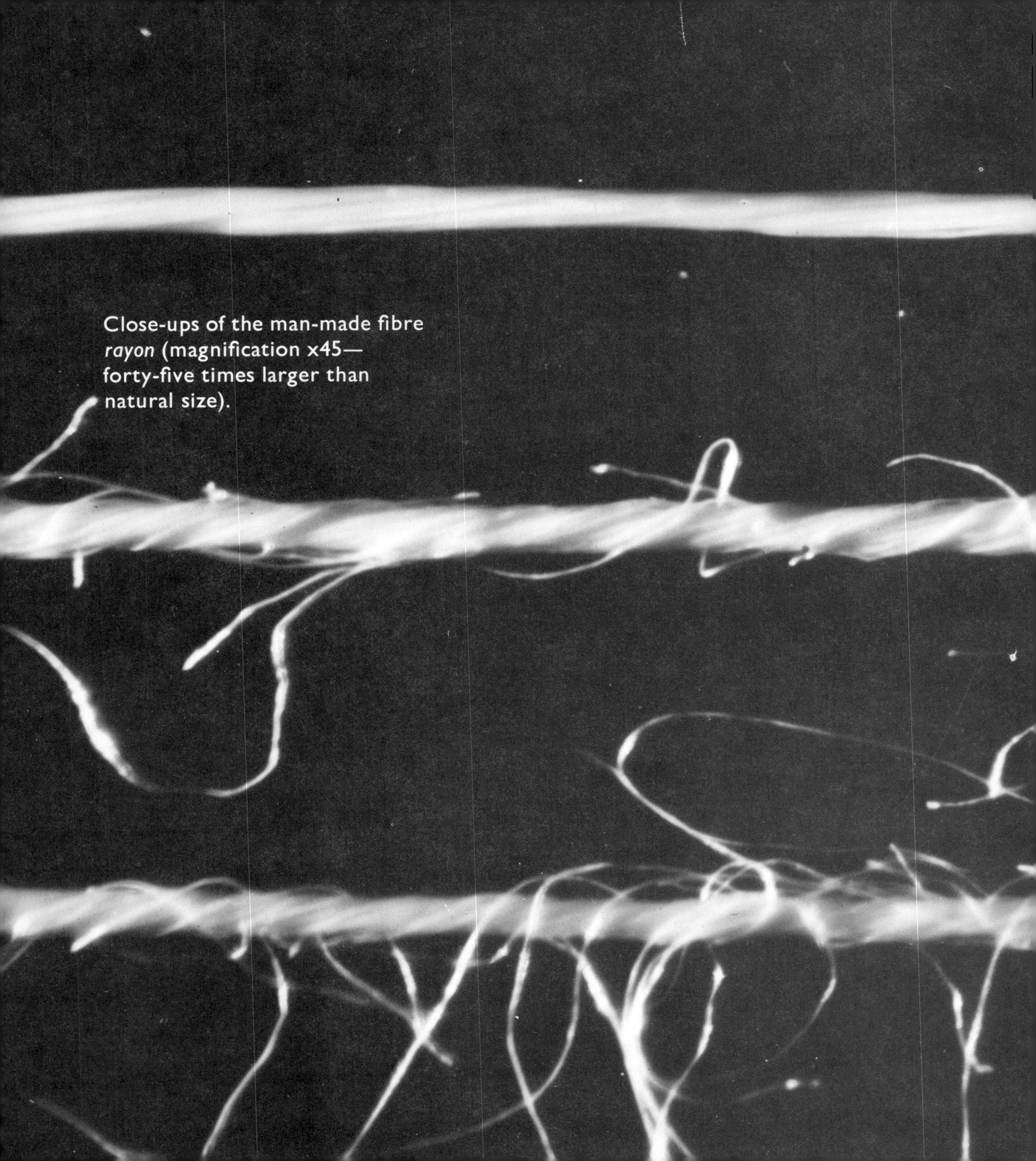

Close-ups of the man-made fibre *rayon* (magnification x45—forty-five times larger than natural size).

Science in Action

Man-made Fibres

Mike Lyth

Priory Press Limited, Hove, England.
Crane, Russak & Company, Inc., New York.

Other books in this series

Science versus Crime William Breckon
Man-Made Fibres Mike Lyth
Lasers William Burroughs
Man's Future in Space Patrick Moore
The Weather Frank Dalton
The Quest for Fuel John A G Thomas
Living with Computers Robert Lewis and Barry Blakeley
Spare Parts for People B J Williams
The War on Pollution Mike Lyth

SBN 85078 212 0

First published in 1975
Second impression 1978
Third impression 1979
Wayland Publishers, 49 Lansdowne Place, Hove, Sussex

Filmset by Keyspools Limited, Golborne, Lancs.
Printed in England at The Pitman Press, Bath

Contents

1 Introducing

... fibres

What do these things have in common?

Nylon ropes.

Look at the things marked by arrows in the pictures. Each of them is made from a quite different material. But all the materials have something in common. Can you think what it might be? If you have no idea at all the answer is on the next page.

Asbestos suit.

Carbon fibre turbine blades.

Hair.

Carpet

Kesp (synthetic meat).

Nylon/PVC sheeting.

Fibrous materials

All the materials shown in the pictures on pages 8 and 9 are *fibrous materials*—materials made up from *fibres*. To show you what that means, here are some pictures showing fibrous materials through a high powered microscope. Can you sort out which of the pictures fits which of the titles in the list? (Answers on page 29.)

1 Skin
2 Cotton fibres
3 A human hair
4 Paper
5 Man-made carpet fibres
6 Wool fibre

Fibres are long thin pieces of stuff —like whiskers or hairs.

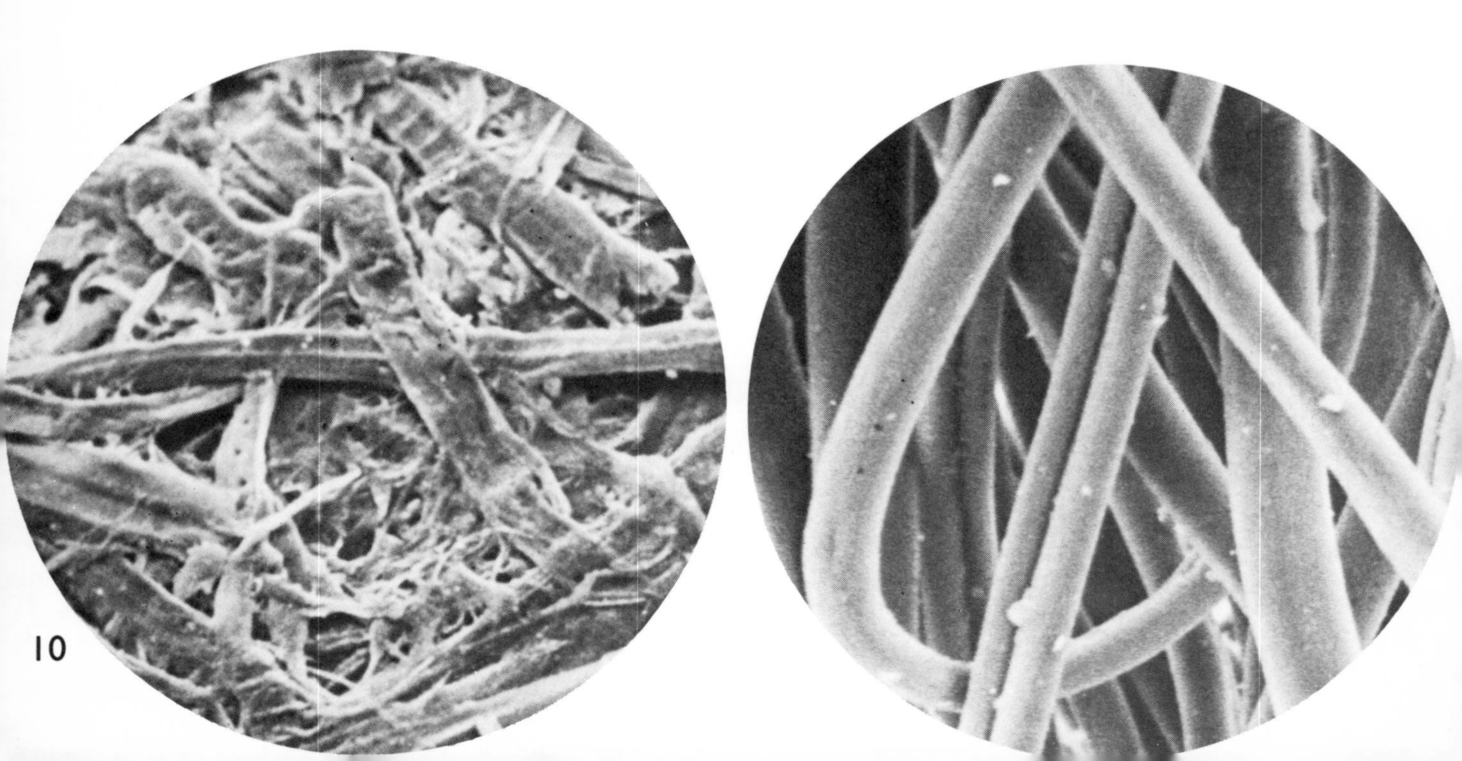

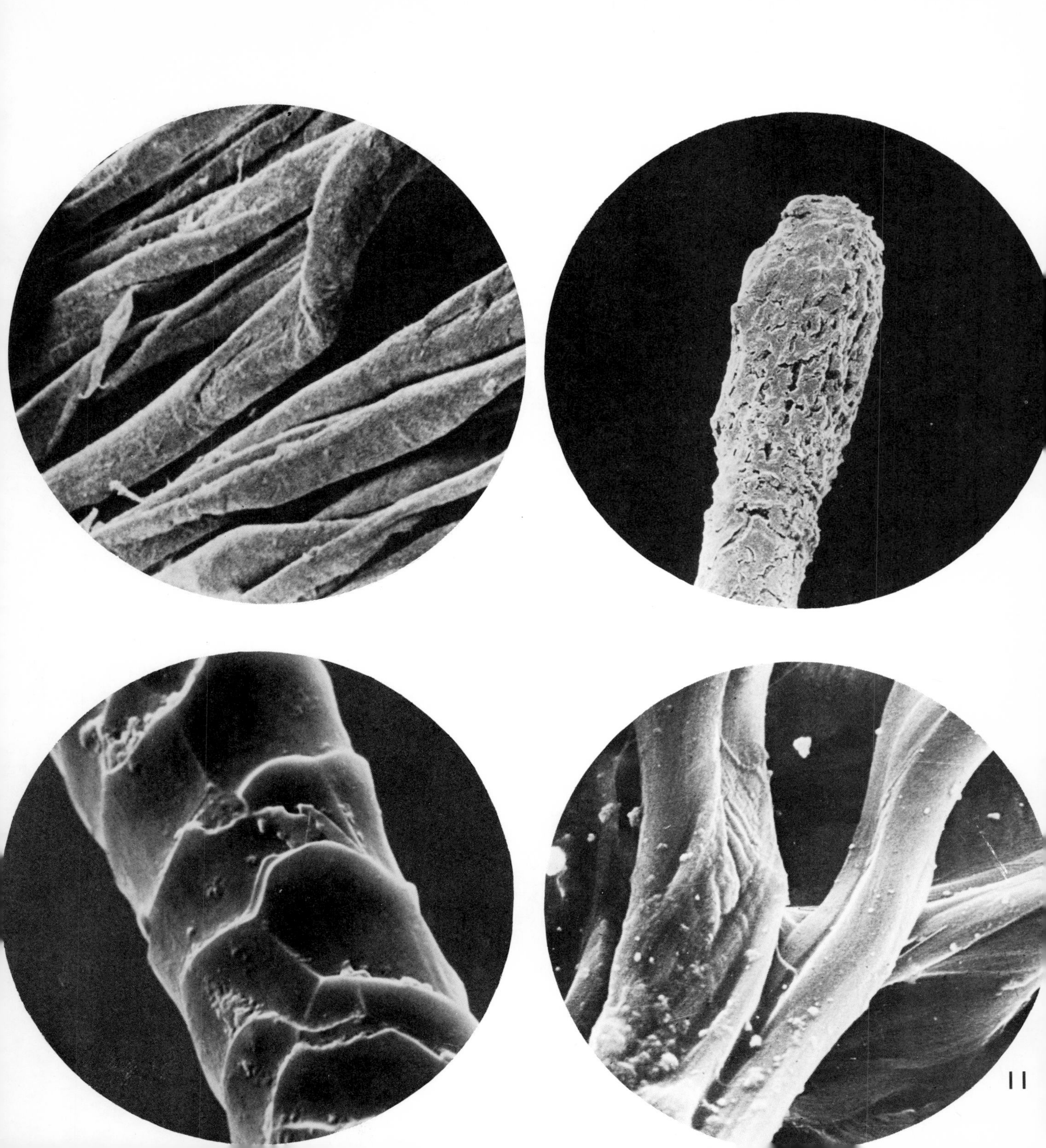

Using natural fibrous materials —from the earliest days of man

Unlike other animals, humans have always been able to protect themselves from bad weather, cold and heat, and shortage of food, by making things. For example, early men made clothes and tents to protect themselves. They made boats with sails to help them move about after food; and they made bows and arrows so that only a few of them could kill a large number of animals. They made most of these things from natural fibrous materials.

Above This sandal is made of twine and is more than nine thousand years old.

Below An encampment in the Sahara desert. Desert nomads like these ones have used natural fibres to clothe themselves and to make tents for thousands of years.

Above The Red Indians of North America used natural fibres to string their bows.

Fibrous materials are light, strong, keep in heat, keep out cold, and are easy to make into useful things.

Left The Vikings wove sails for their longships, and may well have used them to sail to America a thousand years ago.

Where do natural fibres come from?

Some *vegetable fibres* which are made up into useful materials. . . .

Cotton
Flax
Jute
Sisal (from the Agave tree)
Kapok

Agave.

Cotton.

Flax.

Jute.

Kapok.

Squirrel.

Can you find out what each of those fibres is used for? For example, cotton is used for clothing, and sisal is used for ropes. Some *animal fibres* which are made up into useful materials. . . .

Sheep's wool
Silk
Horsehair
Pig's bristle
Squirrel hair

Can you find out what these fibres are used for? Answers on page 29.

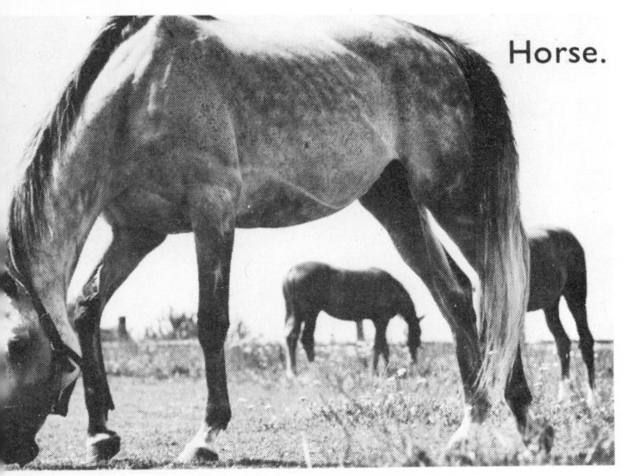
Horse.

Sheep.

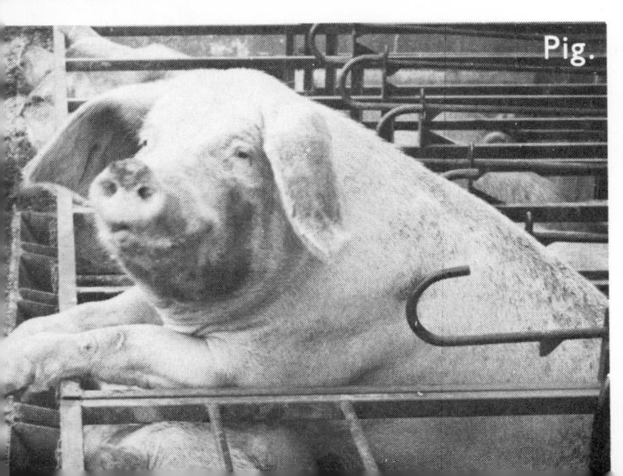
Pig.

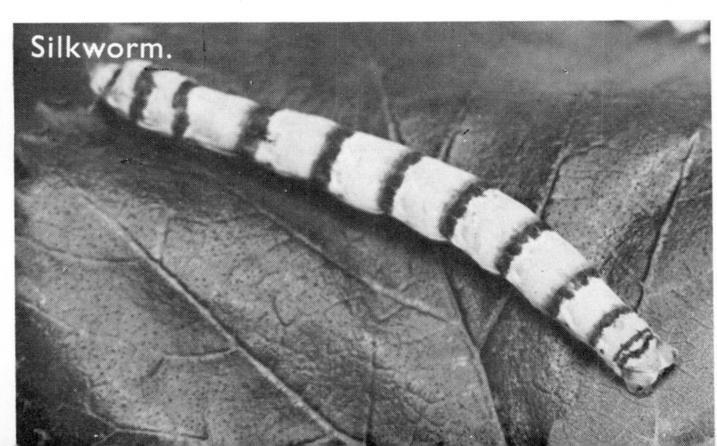
Silkworm.

What about this?
Can you guess what sort of natural fibre we get from this? Hint: it is shown in use on page 8.
The answer is on page 29.

We get our fibres from plants, from animals, and from minerals.

---

2 Fibres and fabrics

"The substance it be out of which the silk-worm wiredraws his clew!"

Left This drawing of silkworms is from Louis Pasteur's book on silkworm diseases. Pasteur started his work on silkworms in 1865.

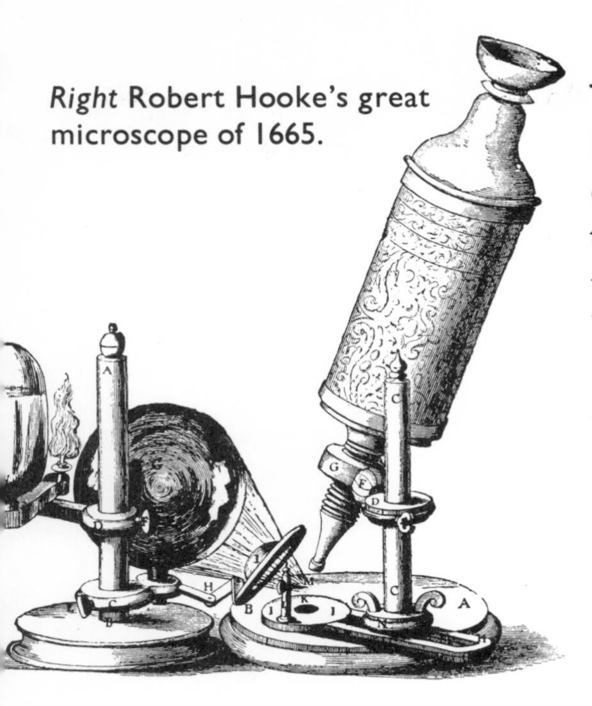

Right Robert Hooke's great microscope of 1665.

The caterpillars in the picture produce a liquid which hardens into a thread when it comes out into the air. They use the threads to make cocoons for themselves—to protect them whilst they change into moths. We call the threads *silk*.

In the days of King Charles the Second, the scientist Robert Hooke (1635–1703) looked at silk fibres under his newly-developed microscope. It seemed to him that if a silkworm could make the liquid from which it spins its thread (Robert Hooke said: "wiredraws his clew") human beings should be able to make it too.

A silk moth leaves the cocoon which it had spun while still a silkworm.

Rayon

More than two hundred years later Sir Joseph Swan (1828–1914) was trying to find a fibre that he could use for the filament of his electric light bulb. He knew that all vegetable fibres are made of a substance called *cellulose*, and he had made threads of it for his experiments. Perhaps he was rather cross to find that his children had taken some of the threads to weave into table mats! That is, until he realized that they had invented a new kind of cloth!

Below Sir Joseph Swan in his laboratory and, to the right, his first electric lamp which he put on display in Newcastle in 1878.

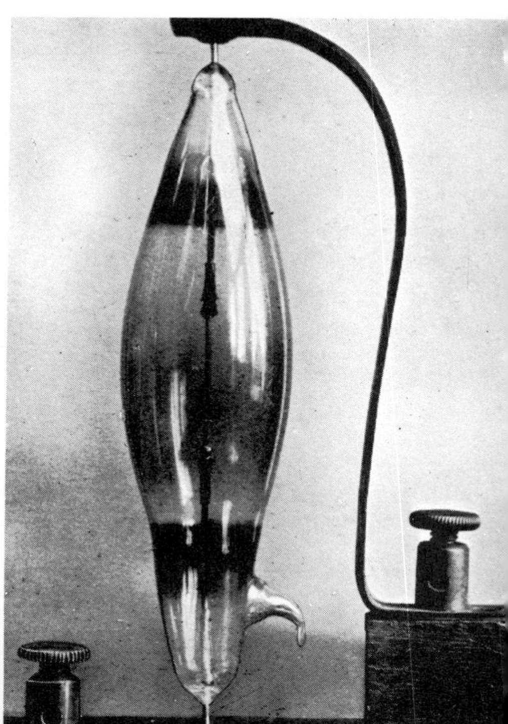

In the middle of the nineteenth century a terrible disease struck the silkworms in France. It raged for twelve years and almost wiped out the silk industry. The great scientist Louis Pasteur (1822–95) did much to find a cure. One of the scientists working with him—Count Hilaire de Chardonnet (1839–1924)—started to think about finding a man-made replacement for silk in case all the silkworms died. He put together the ideas of many earlier scientists and in 1889 opened the first man-made textiles factory. Count Hilaire de Chardonnet called his man-made fibre "artificial silk." Today we call it *rayon*. The story of rayon production starts with a tree. . .

Right Louis Pasteur in 1881, proudly wearing the Ribbon of the Grand Cross of the Legion of Honour. This was one of the many honours he received for his scientific achievements.

If you looked at a thin slice from a tree, under a microscope, you would see that it was divided up into little boxes. The little boxes are called *cells*. The walls that divide one cell from the next are made of the substance called *cellulose*. Rayon fibres are made of cellulose too. To make rayon the

cellulose has to be extracted from the wood, then turned back into a fibre again.

1. To start with, the wood is sawed and ground up into very tiny bits. Mixed with water, it is called wood pulp.

2. Then it is soaked in a chemical called caustic soda to untangle the very very tiny fibres that the bits are made up of.

A "biscuit" of wood pulp coming out of a press in Swaziland.

Sheets of wood pulp being slipped into baths of strong caustic soda solution.

3. Another chemical called carbon disulphide is added. This changes the mixture from a whitish sloppy mess into solid yellow crumbs. This process is called *xanthation*.

4. These are left to soak for several days in another caustic soda bath. If the temperature and the dryness of the air near the bath are just right, the crumbs change into an orange syrupy liquid called *viscose*.

5. Finally the viscose is squirted into a bath of another chemical—sulphuric acid. As it squirts through the nozzle it changes into long thin fibres of pure cellulose.

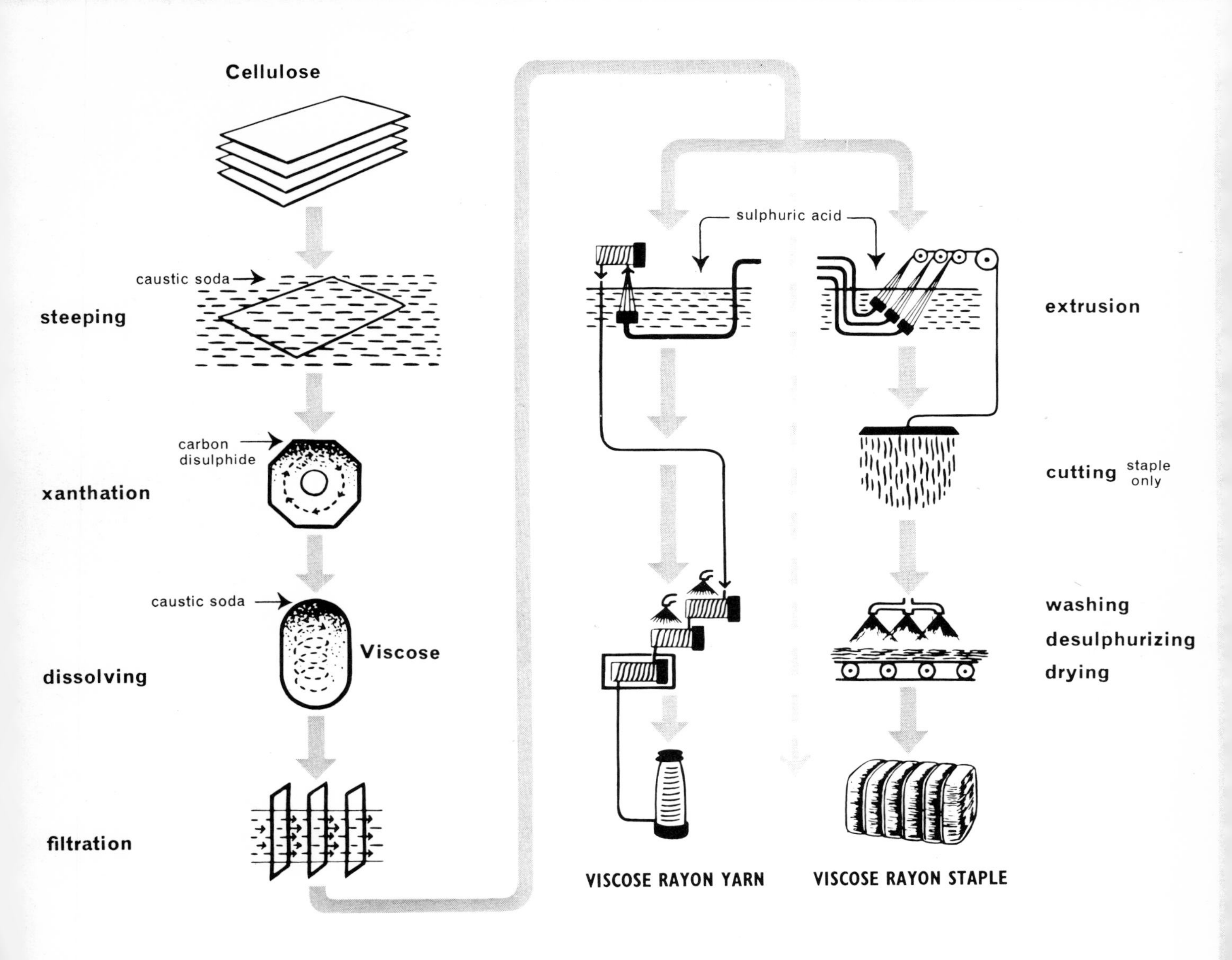

Here is a chart showing all the stages of rayon production. The cellulose fibres can be made into *yarn* or *staple*. More about this on page 36.

Viscose rayon has been used to make tassels, braids, knitted ties, sweaters, and carpets. It was found that thicker "chunky" fabrics could be made by chopping up the long fibres into little bits, then spinning them together into a thread. This is the same way that we use the short fibres of cotton and wool. The only problem with rayon is that it all starts with trees, and trees take time to grow.

Answers to questions

Answers to question on page 10

C D

A B E F

A. Paper (Magnification x350)
B. Man-made carpet fibres (Magnification x264)
C. Cotton fibres (Magnification x800)
D. A human hair (Magnification x225)
E. Wool fibre (Magnification x1,560)
F. Skin (Magnification x1,800)

Answers to questions on page 15

Here are some of the uses of the natural fibres mentioned.
Cotton: cloth and clothing, from heavy sailcloth, towels and bandages to lightweight shirts.
Flax: linen for clothing, table cloths, canvas and fish nets.
Jute: carpets, mats and ropes.
Sisal: ropes, mats and brushes.
Kapok: stuffing for furniture, pillows, sleeping bags and lifejackets.
Wool: cloth and clothing, from blankets to woolly socks.
Silk: luxury clothing.
Horsehair: stuffing for mattresses and furniture; all kinds of brush.
Pig's bristle: paint brushes, clothes brushes and many other types of brush.
Squirrel hair: fine paint brushes, expensive fur coats and hats.

Answers to question on page 16

The photograph on page 16 shows a special kind of rock called magnesium silicate. It is made into asbestos fibres—for fire-fighting suits for example.

Answers to question on pages 50 and 51

1	2	3
4	5	6

1. Reinforcing a polyester resin with glass fibres.
2. Insulating an attic against heat loss with glassfibre wool.
3. Under bonnet sound insulation in a car.
4. Glassfibre pad for filtering liquids. In this case the liquid is a hospital patient's blood!
5. Glass fibre woven into curtain material.
6. Making a glassfibre reinforced plastic moulding of a coat of arms.

What's so special about cellulose?

Scientists imagine that everything, including ourselves, is made up of tiny little objects which they call *atoms*. These atoms are so small that you can't see them, even under the most powerful microscope. You could line up ten million of them across a full stop on this page. Substances we know very well, like the oxygen gas we breathe, or water, or cellulose, are made up of little groups of atoms. These groups of atoms are called *molecules*. Magnified millions of millions of times some molecules might look like this.

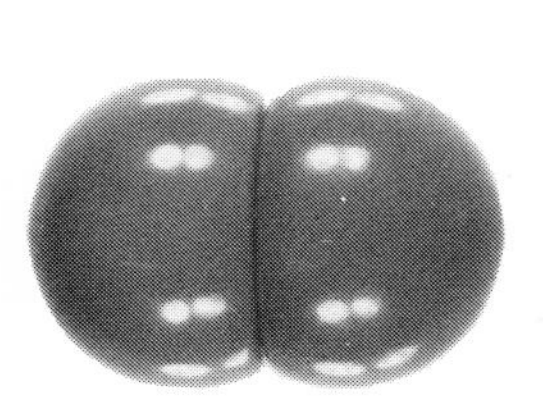

The one on the left is a molecule of oxygen made of two atoms of oxygen. The middle one is a water molecule. The right hand one is a molecule of a gas called ethylene. In the world of molecules these are rather small ones. Some are very much bigger.

Have you used a plastic bag today? If you have, the chances are that it was made of *polythene*. If you wanted to make a model of a polythene molecule it would look like this.

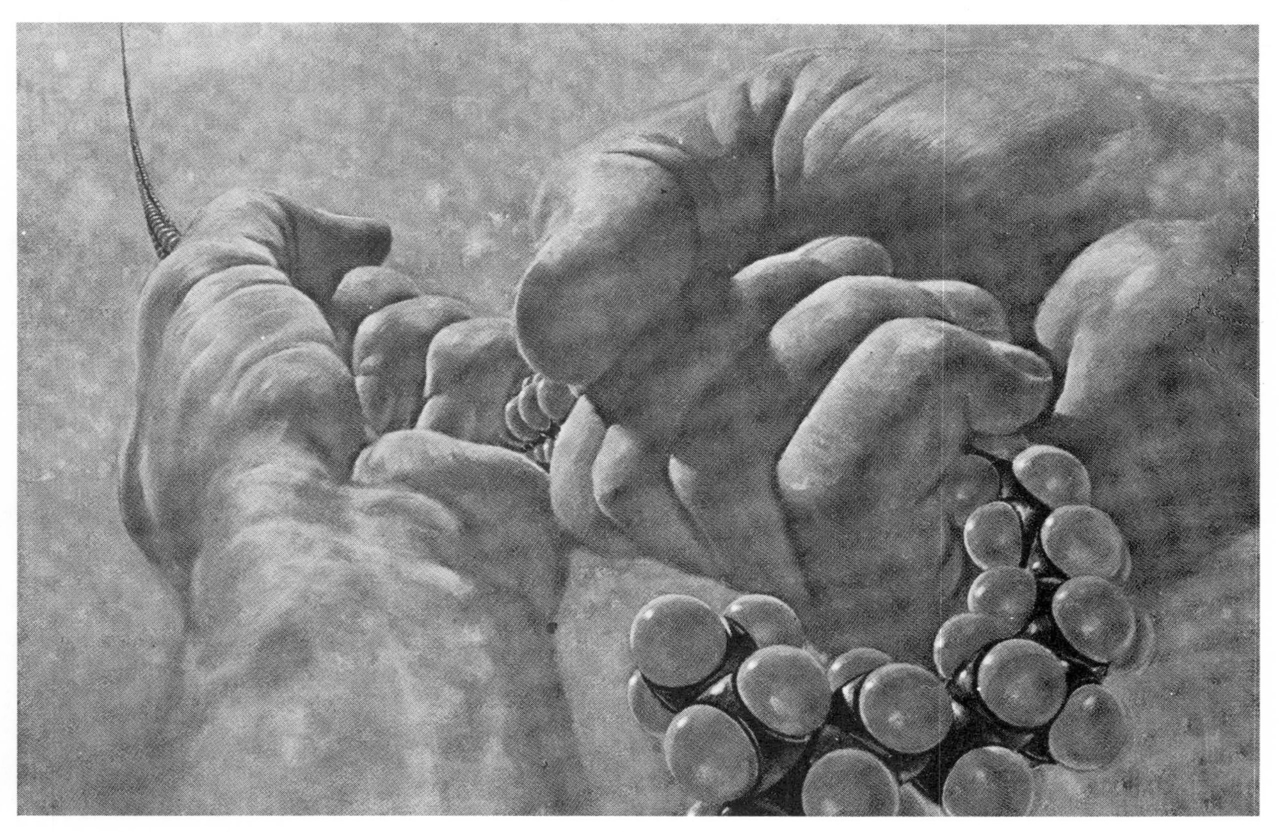

Above A model of a polythene molecule, *millions* of times bigger than a real polythene molecule.

Substances, like polythene, which have their atoms and molecules connected together to make long, long, chains are called *polymers*. Cellulose is a natural polymer. Most of the fibres we turn into fabrics are made from polymers.

Improving on nature?

About fifty years ago scientists began to look for ways of making polymer fibres without using wood or other natural substances. The idea was to get two chemicals to react together so that their molecules joined up into long chains. The difficult part was finding the two chemicals to start with. In February 1928 a team of scientists led by Dr. Wallace Carothers (1896–1937) started doing research into the problem at the Du Pont Chemical Company in the USA.

Seven years later they had the answer. The two chemicals used (*hexamethylene diamine* and *adipic acid*) could be extracted from coal or oil. They named the new material *nylon*. It turned out to feel very much like silk and wool. Also it was enormously strong, even when made very thin indeed. It was springy, and hard wearing. Very soon, every lady in the country was crying out for the beautiful new stockings made from it.

Nowadays stockings and tights are still on sale, but lots of other uses have been found for nylon.

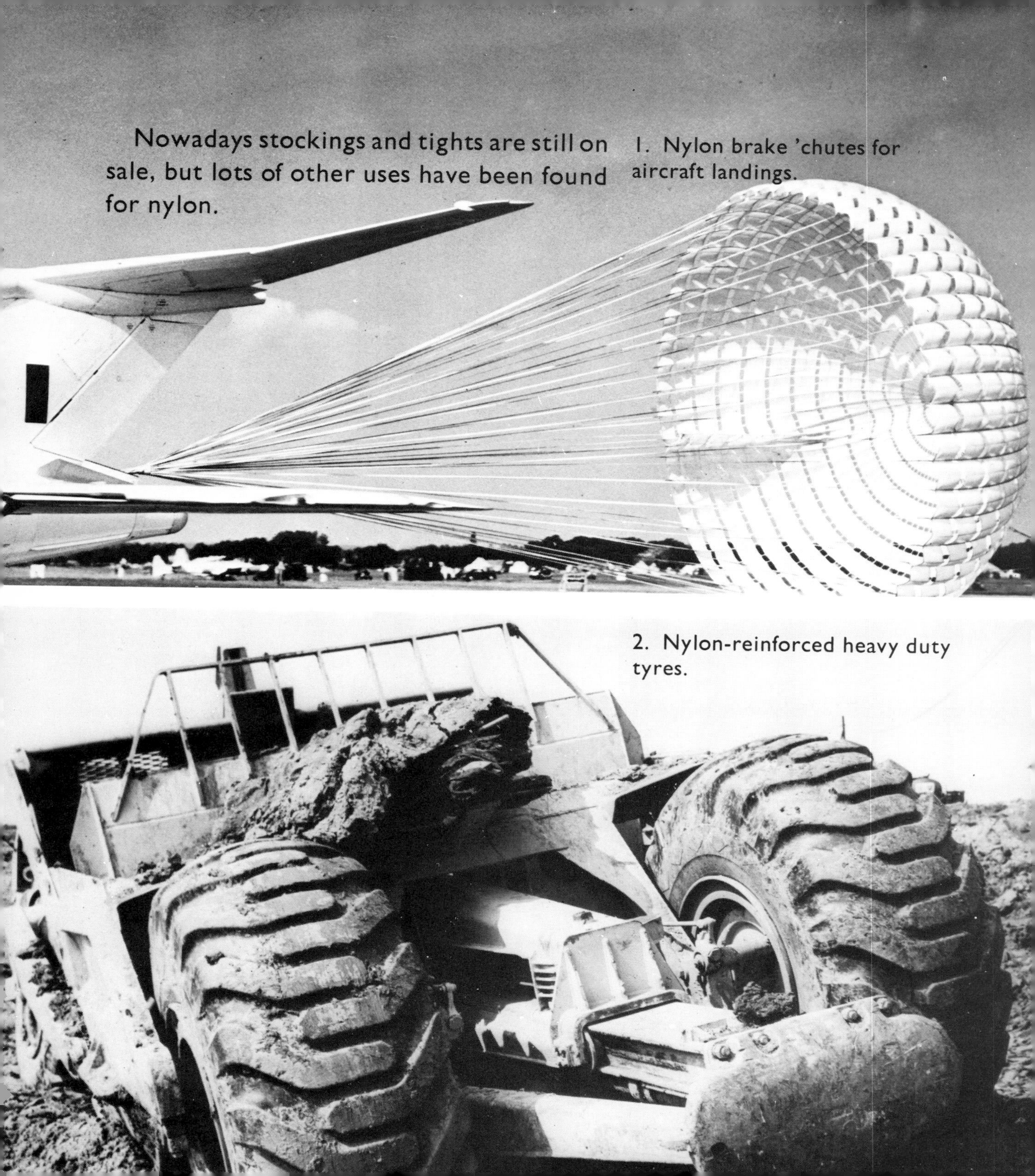

1. Nylon brake 'chutes for aircraft landings.

2. Nylon-reinforced heavy duty tyres.

Throw-away lighters.

4. Mountaineering ropes.

5. Safety nets for bridge paint

6. Warehouse "air-house" roof—kept up by powerful air fans.

Fibres into fabrics

Whether the fibres are natural or man-made, a lot of work has to be done to change them into fabrics. This is what happens.

Natural or man-made fibres are made from atoms and molecules far too small to see.

Full length man-made fibres are spun into *continuous yarn*. Chopped up man-made fibres (called *staple* fibres) or natural staple fibres are spun into "hairier" *staple yarn*.

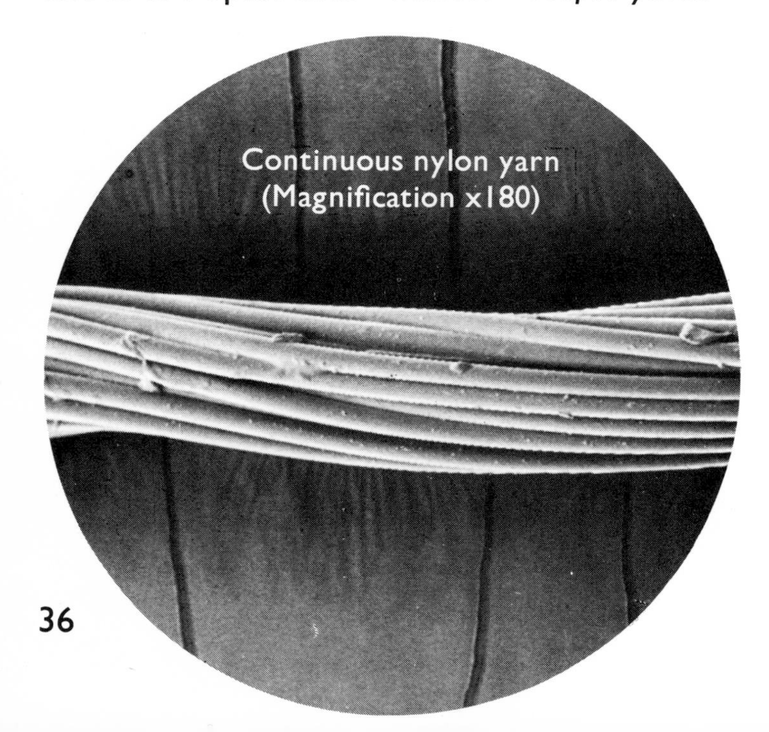

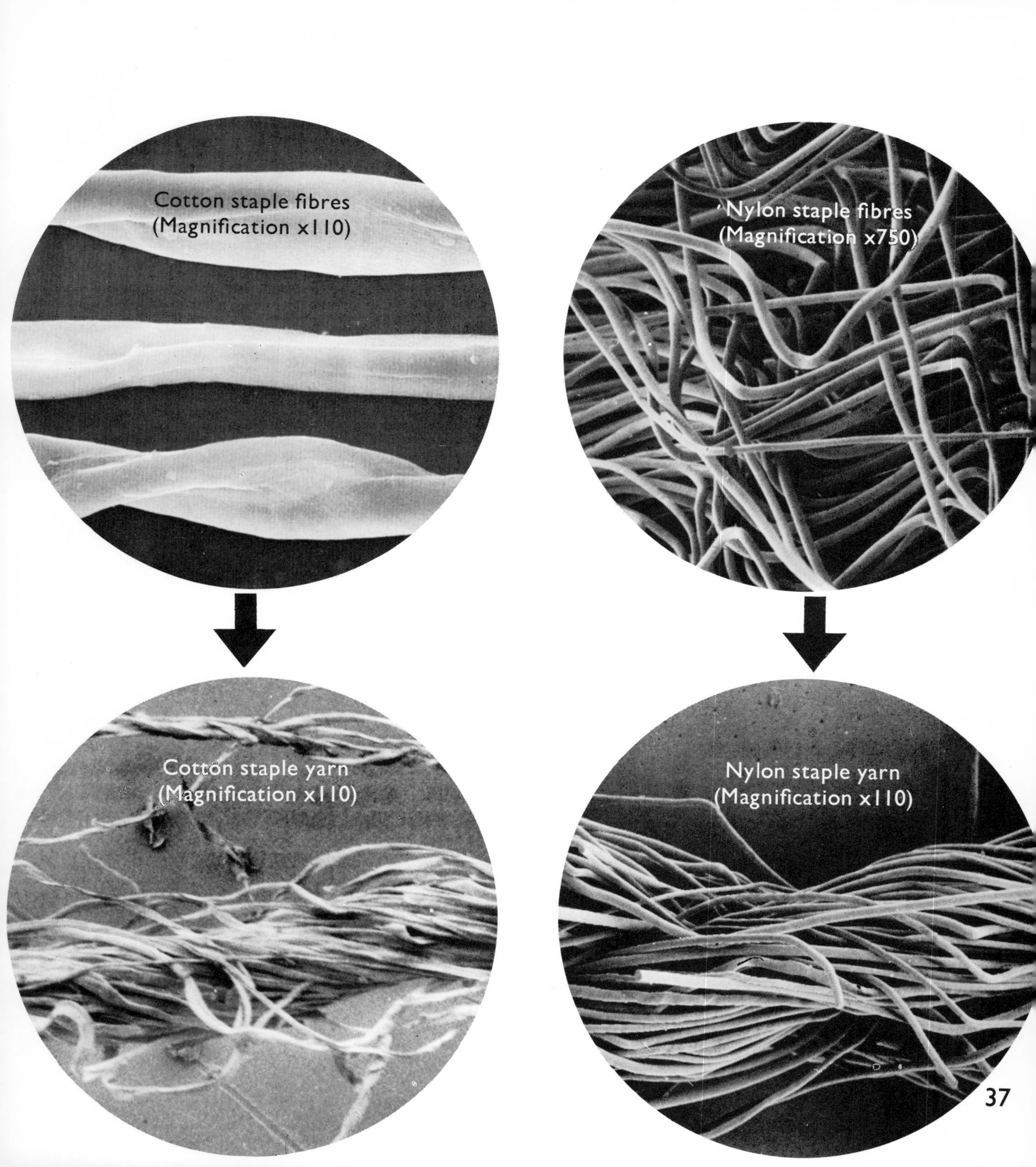
Cotton staple fibres
(Magnification x110)
Nylon staple fibres
(Magnification x750)
Cotton staple yarn
(Magnification x110)
Nylon staple yarn
(Magnification x110)

Fabrics are woven from yarn. *Knitted fabrics* are made by tying one strand of yarn to the next with a line of knots. *Woven fabrics* are made from many separate strands, with at least two sets running across each other. The strands lock each other into place.

Close-up of a knitted fabric.

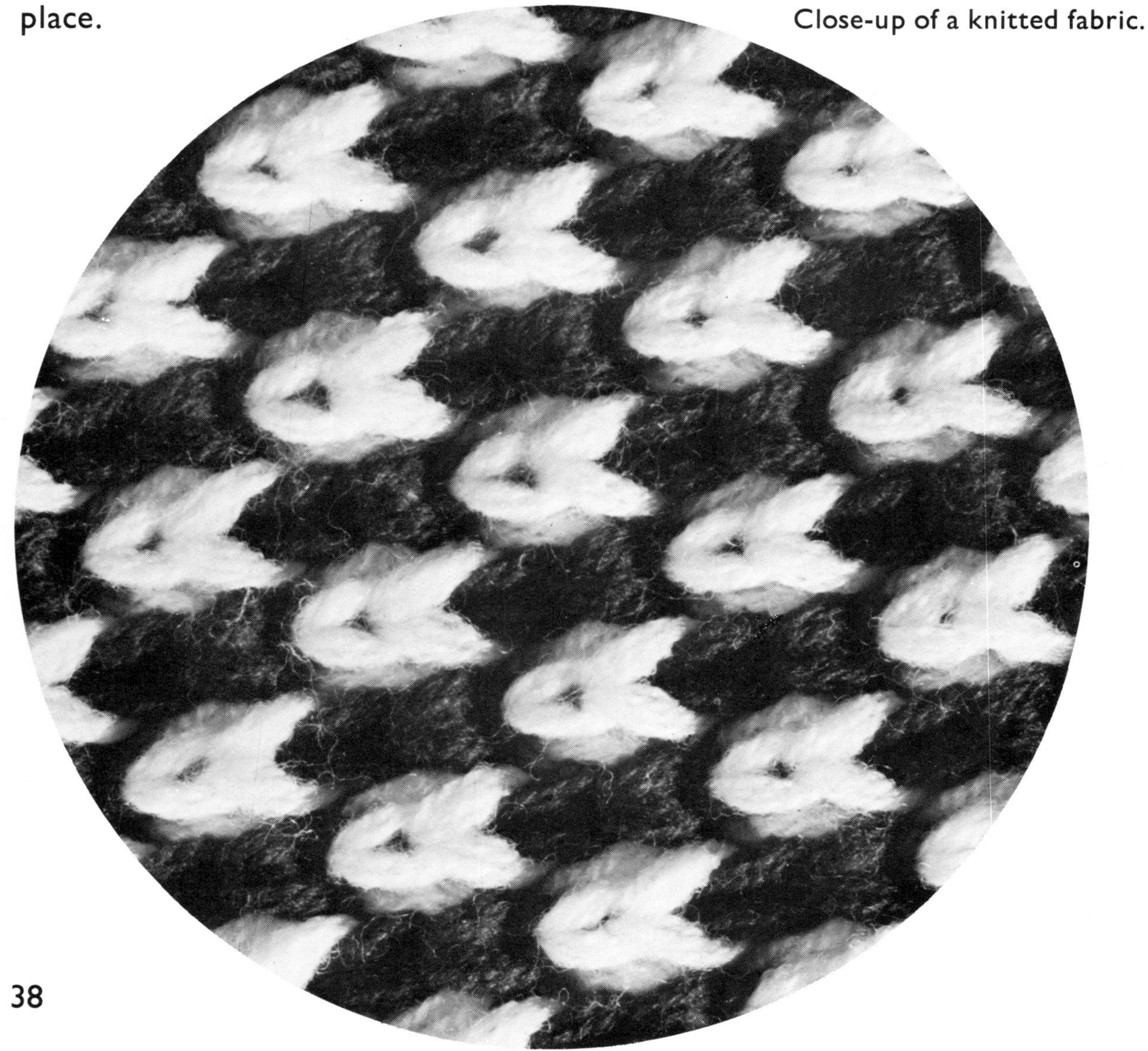

Close-up of a woven fabric.

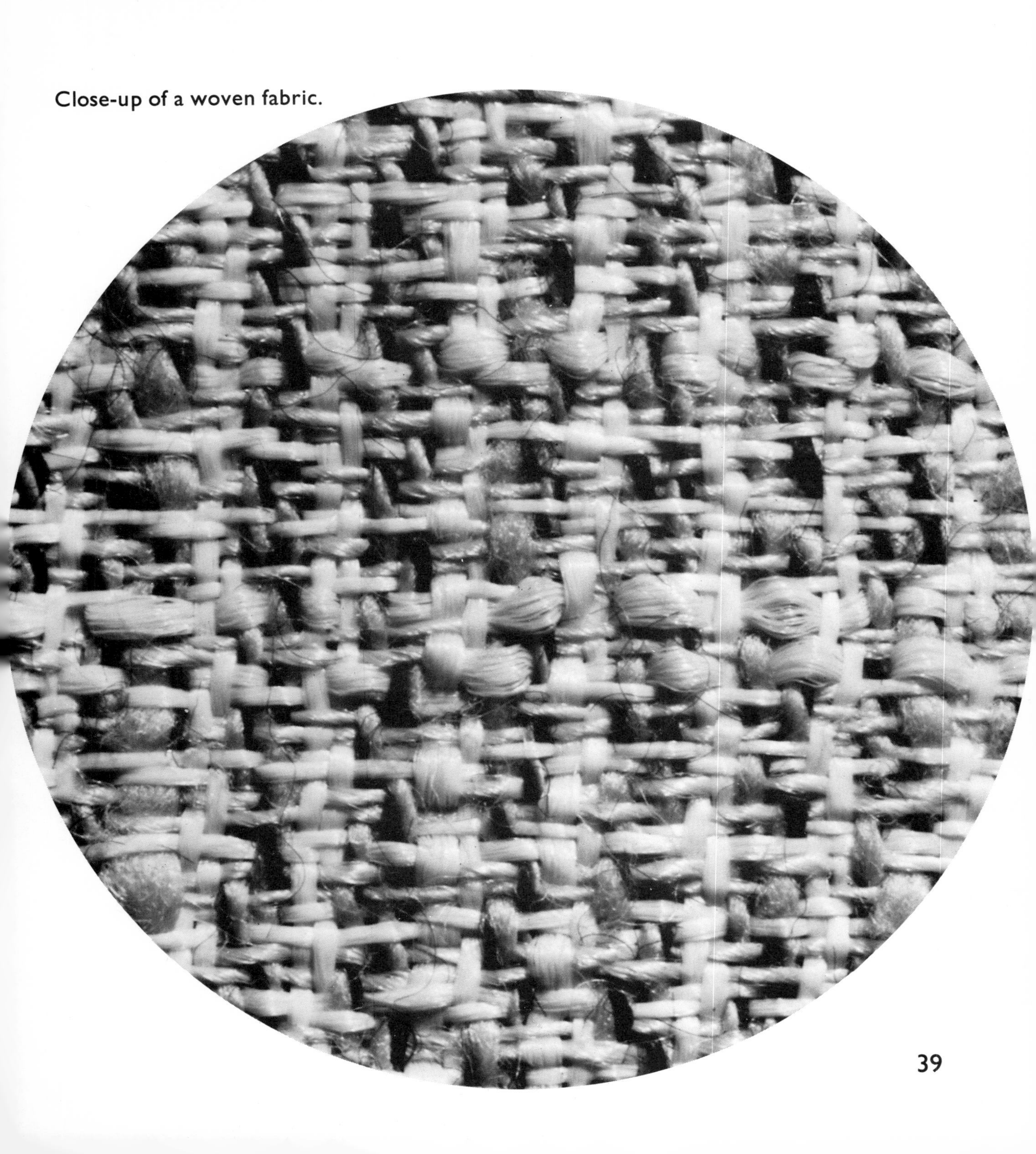

Clothes are cut and sewn from fabrics.

Fabrics galore

Today's clothing designers have a very wide choice of fabrics. With natural fibres, you can make different fabrics by spinning different thicknesses of yarn, by mixing different yarns, by dyeing in different colours, and by choosing different knitting and weaving patterns. You can do all these things and many more with man-made fibres.

You can leave the fibres smooth and straight or give them all sorts of curls and crimps.

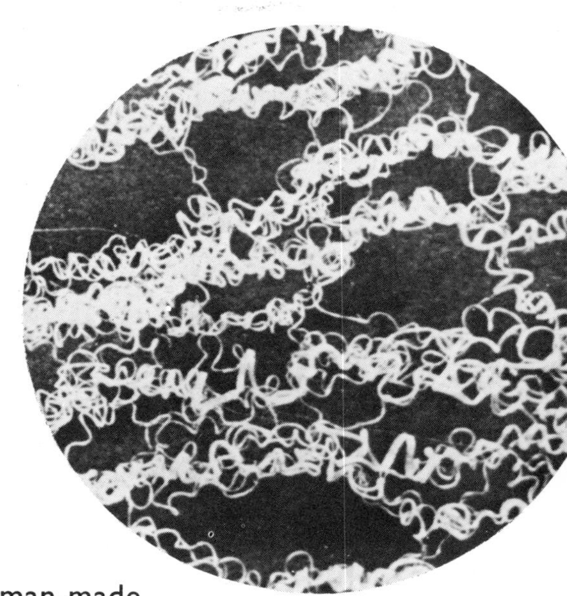

Close-ups of the same man-made fibre after different treatments.

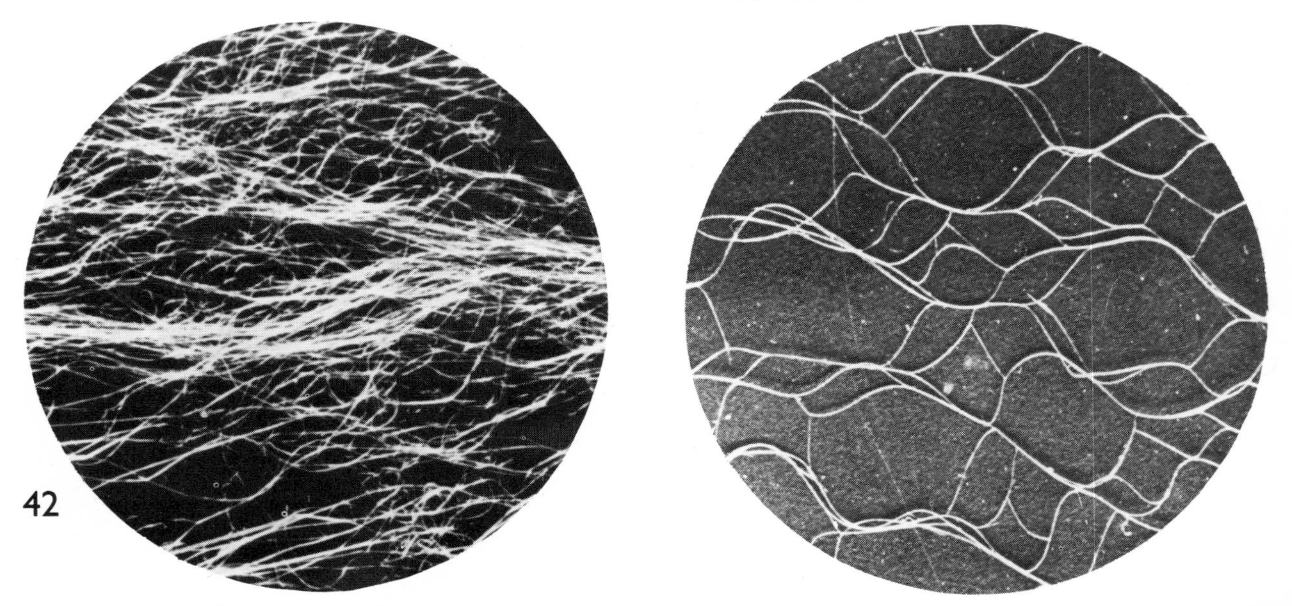

Smooth and shiny . . .

Fabrics made from smooth fibre can be shinier, smoother, closer fitting, harder wearing.

Fabrics made from curly fibres can be thicker, fluffier, warmer, more stretchy.

. . . thick and chunky.

Most man-made fibres will melt if you heat them and set again when they cool. They are called *thermoplastic* materials—or just "plastics" for this reason. You can use this property to cut them out, and to form them into unusual shapes. And you can use it to make fabrics without doing any knitting or weaving at all.

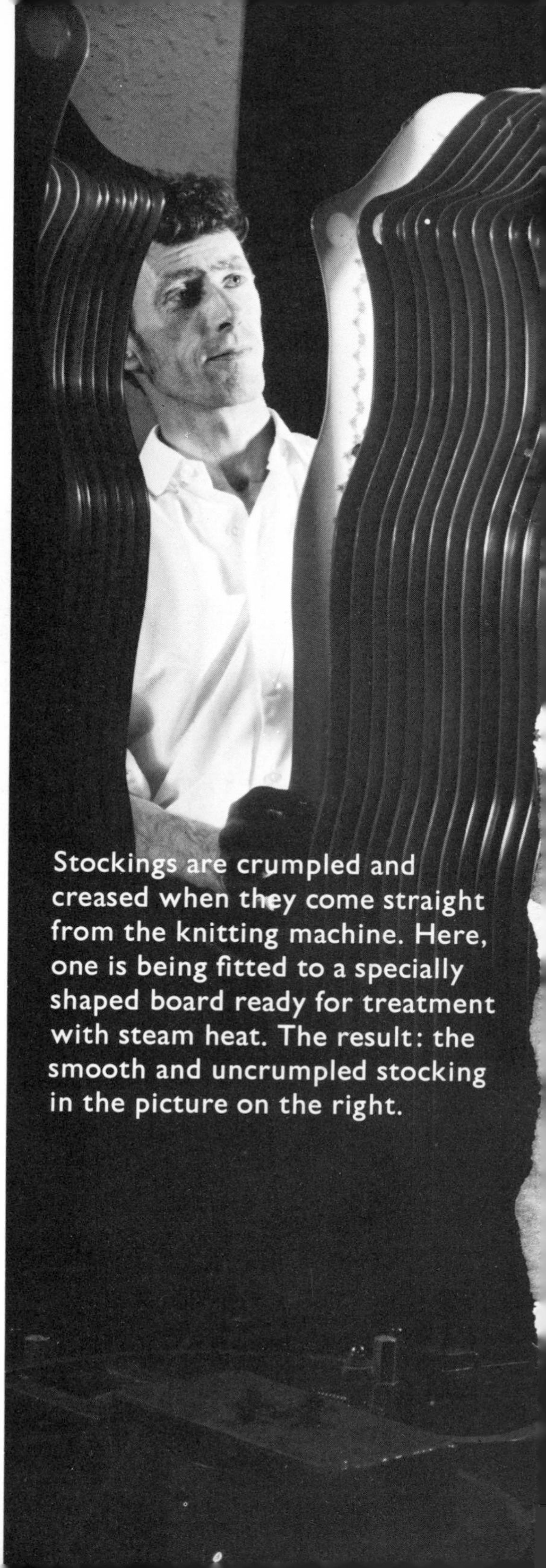

Stockings are crumpled and creased when they come straight from the knitting machine. Here, one is being fitted to a specially shaped board ready for treatment with steam heat. The result: the smooth and uncrumpled stocking in the picture on the right.

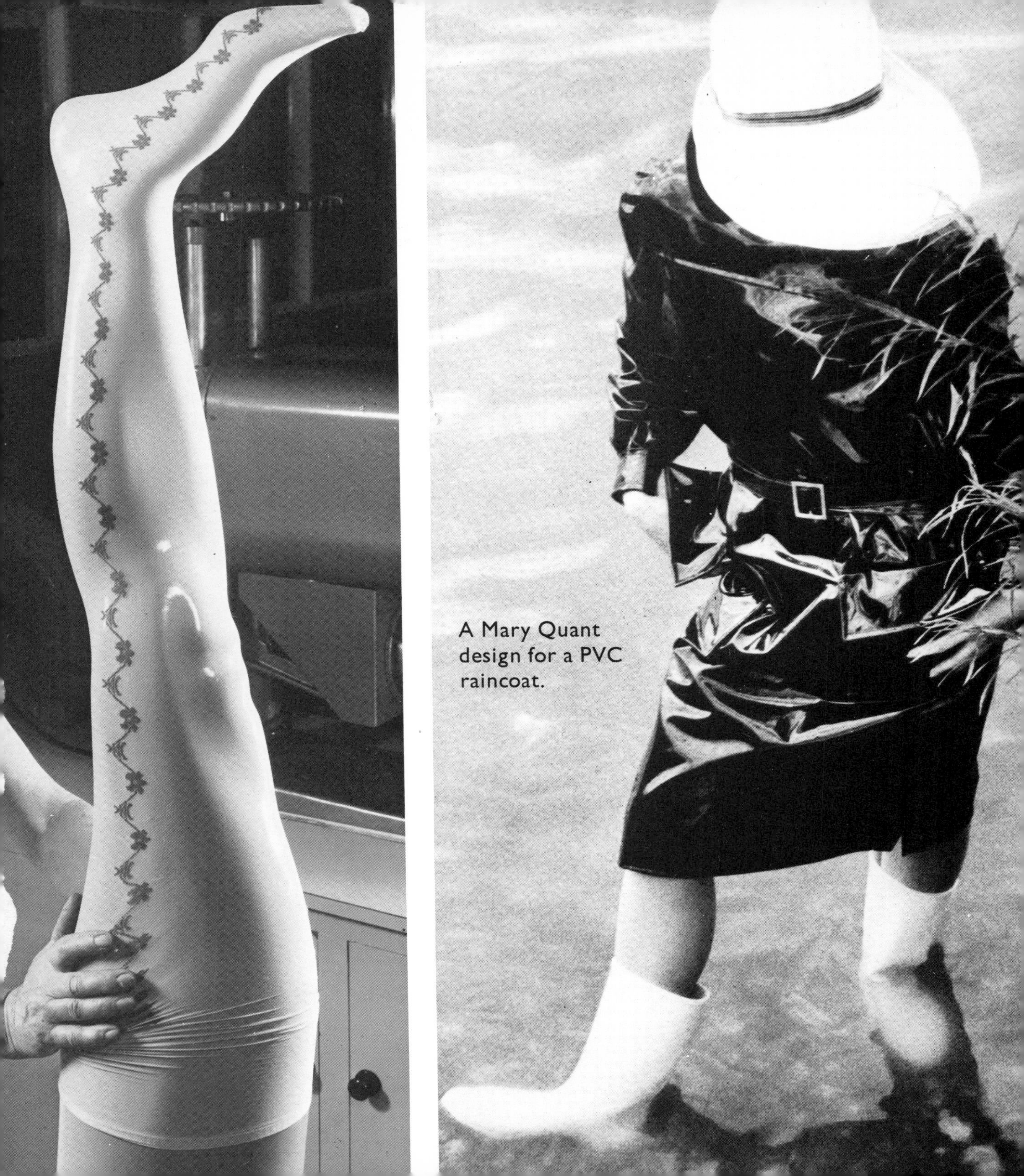

A Mary Quant design for a PVC raincoat.

New uses for man-made fibres

Every day someone thinks of a new use for polymer fibres—other than in clothing designs. A lot of the ideas turn out to be unworkable. Here are some that should be very workable and very useful.

A man-made heart valve woven from polymer fibres works just as well as the real thing—outside the patient's body. The problem is—will it work inside, or will the heart's tissues reject it?

An "instant dam" ready for use anywhere in a few minutes. Made from man-made

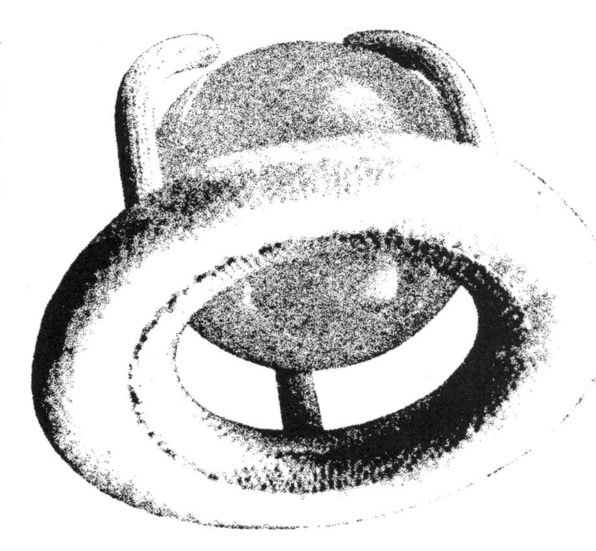

Above A man-made fibre heart valve. It works like the ping-pong ball in the end of a snorkel. As blood pushes upward it flows easily through the ring and round the ball. But if the blood tries to flow downwards, the ball is jammed tight into the hole.

Left An "instant dam" made of man-made "rubber nylon."

Above A hovercraft with a man-made fibre skirt.

rubber and nylon, it is filled with water from the stream itself.

One of the big problems in designing a hovercraft was how to get the right sort of "skirt." It had to hold in the "air cushion" without dragging too much on the water. Most "skirts" today are made from man-made fibres.

Thrown into the sea, the "man-made sea-weed" fibres on the mat below collect sand around themselves. The result is an under-water sandbank just where it is needed.

Right Man-made fibre seaweed.

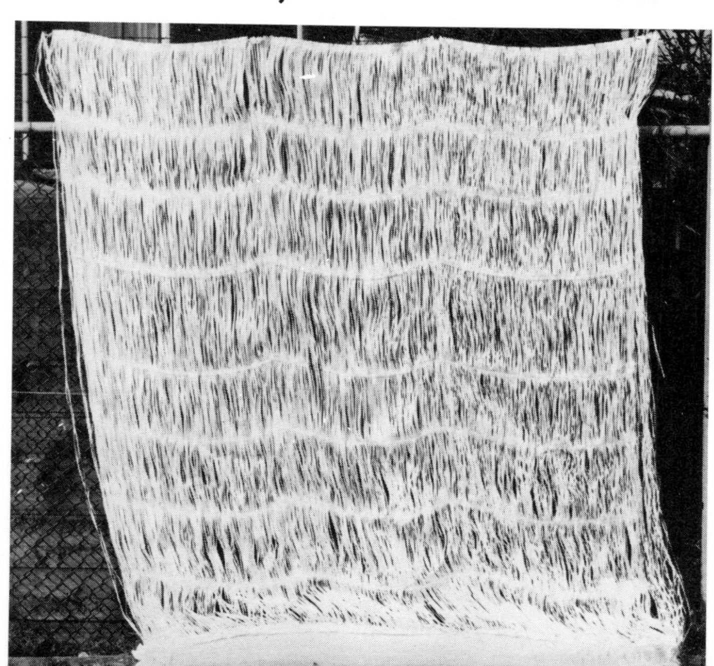

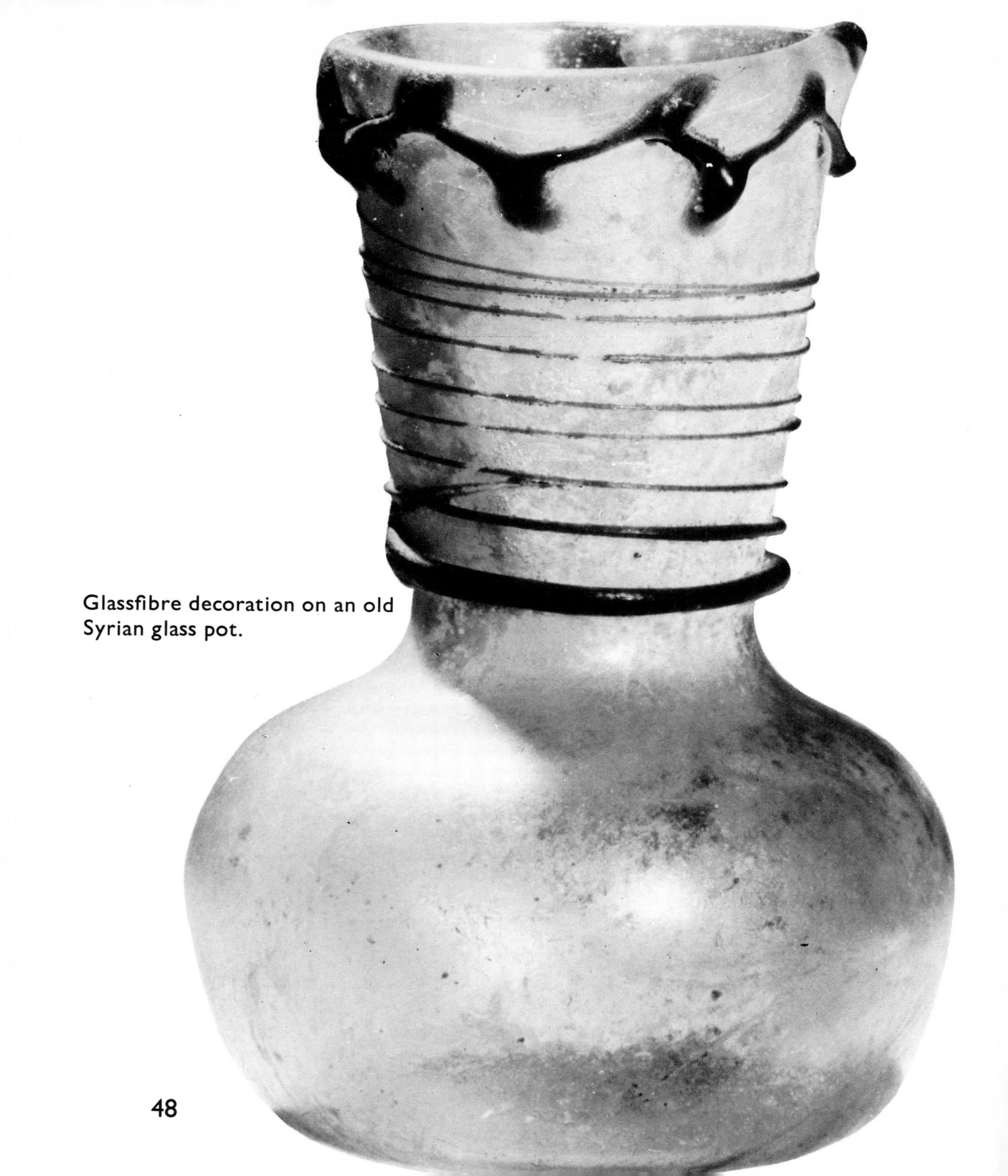

Glassfibre decoration on an old Syrian glass pot.

3 See-through fibres

Glass fibres

Hundreds of years ago glassworkers melted glass and drew it out into fine fibres to decorate their ornaments. We still use glass fibres as decoration today—usually in ways that would have surprised the early workers! One example is their use in this table lamp below.

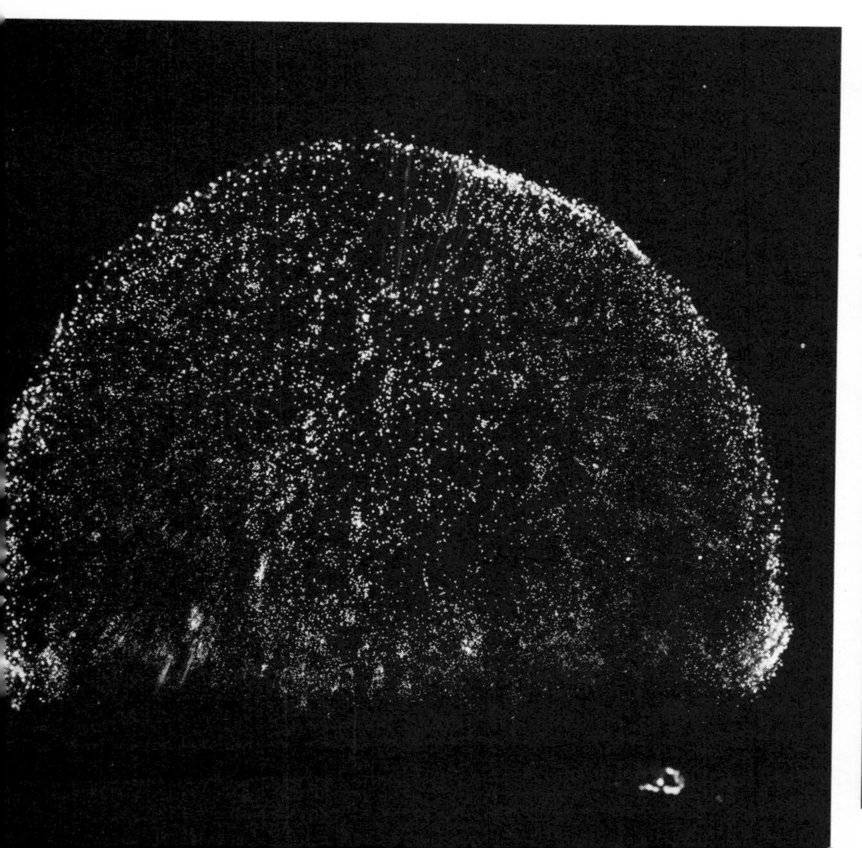

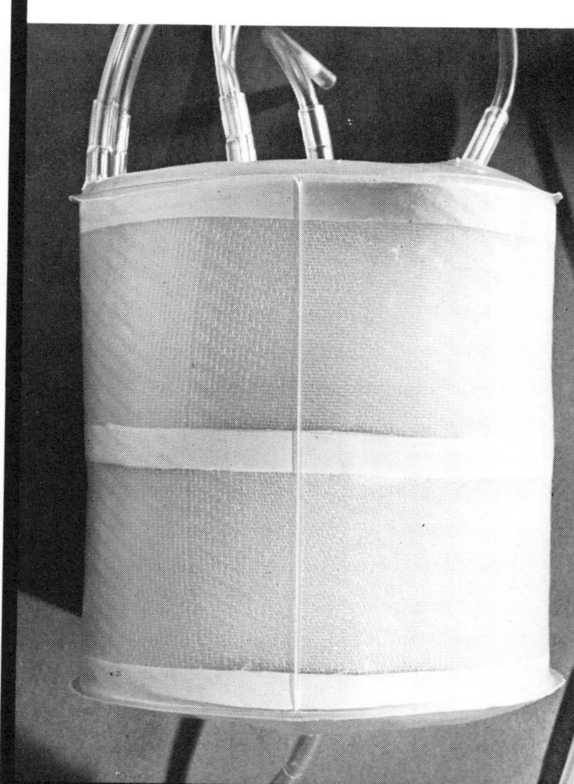

We have found lots of other uses for glass fibres. Can you say what they are being used for in the pictures here? The answers are on page 29.

How do they make glass fibre?

Here are two of the ways of making glass fibre.

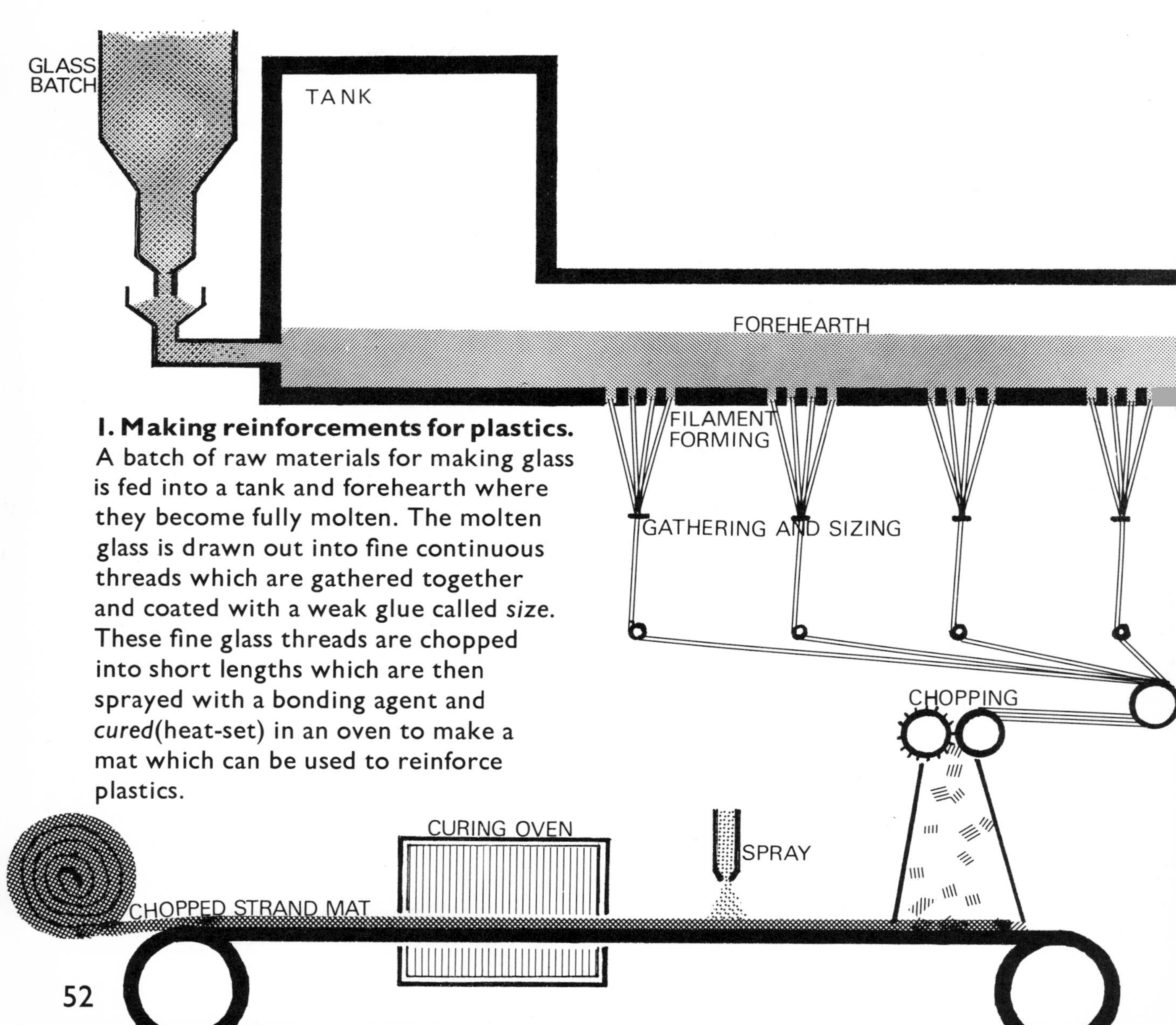

1. Making reinforcements for plastics. A batch of raw materials for making glass is fed into a tank and forehearth where they become fully molten. The molten glass is drawn out into fine continuous threads which are gathered together and coated with a weak glue called *size*. These fine glass threads are chopped into short lengths which are then sprayed with a bonding agent and *cured*(heat-set) in an oven to make a mat which can be used to reinforce plastics.

2. Making insulation material. A thick stream of molten glass flows from the forehearth into a rapidly spinning steel dish with hundreds of very small holes in its sides. Glass in the spinner is thrown through these tiny holes and forms a mass of thin fibres. These fibres are blasted with hot gas, sprayed with a bonding agent and drawn onto a moving conveyor belt through a curing oven. The glassfibre wool mattress is then cut, trimmed and rolled up into suitable lengths.

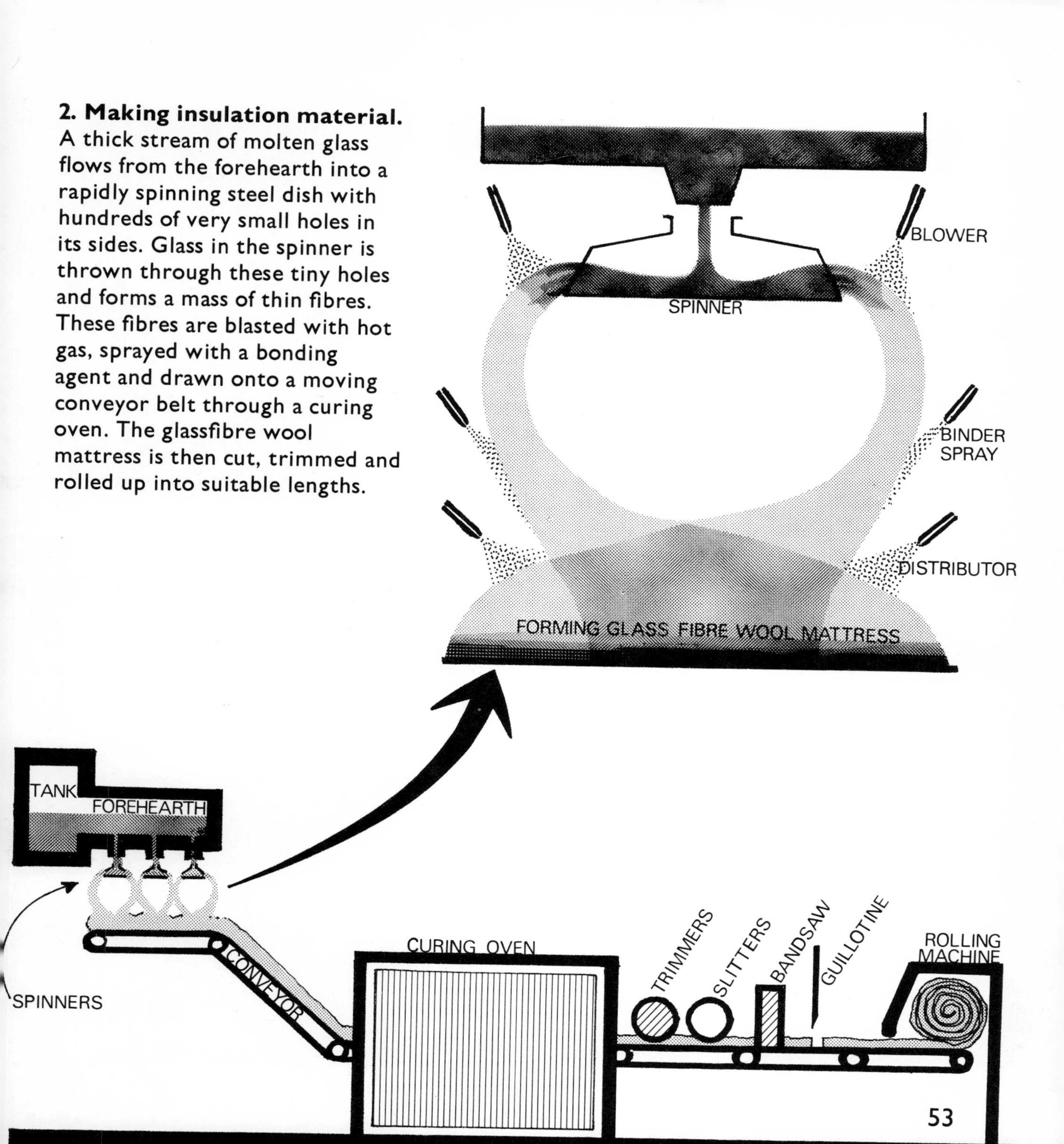

Fibre optics

For many years scientists have known that beams of light will bounce along inside a glass rod or tube. If the beam is going in just the right direction it is "trapped" inside the rod. This can even happen if the rod is bent. The thinner the glass rod the more it can be bent before the light "escapes."

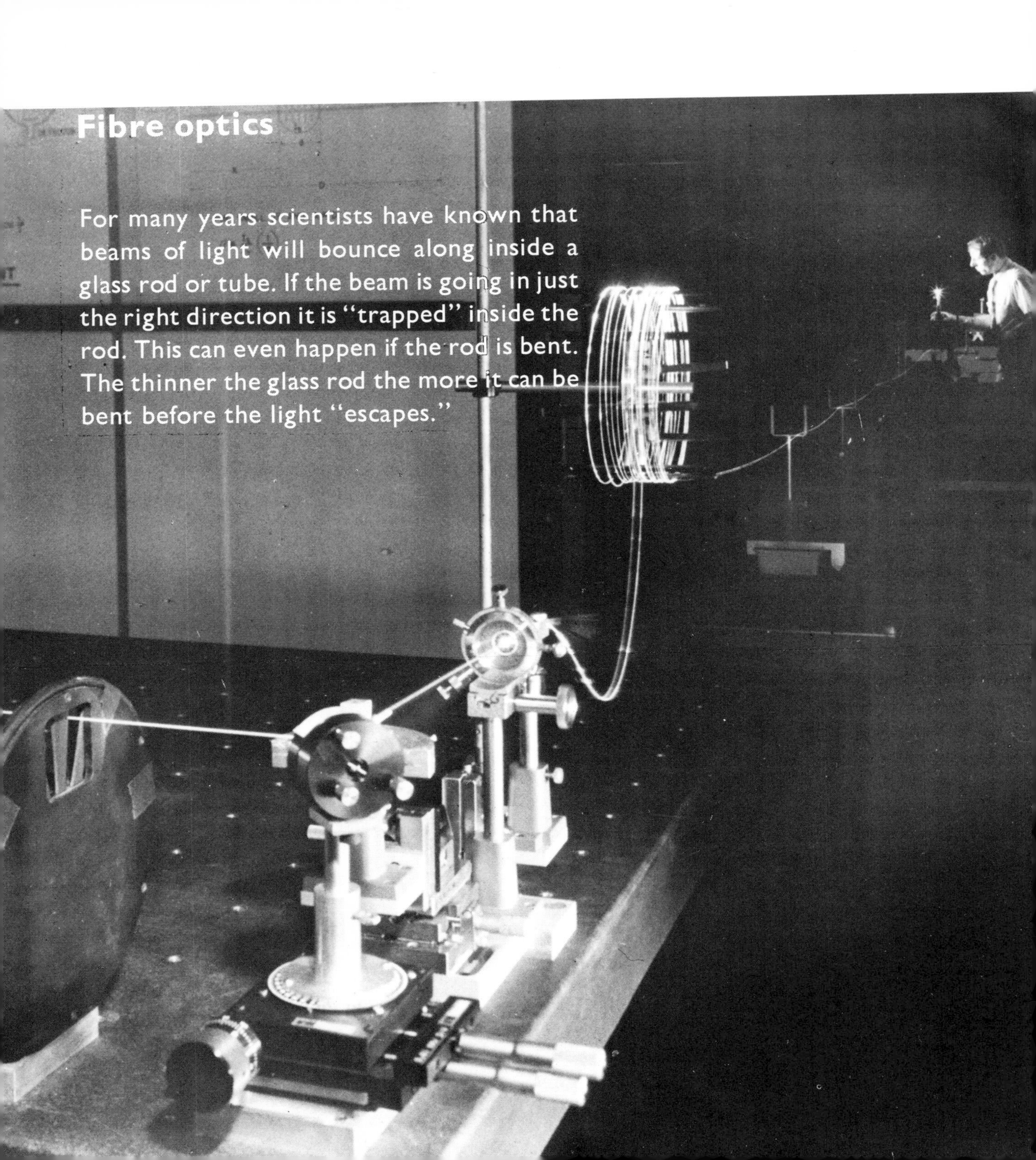

This idea has led to one of our most exciting inventions—*fibre optics*.

Each of the "dots" in the motorway signs in the picture is at the end of a bundle of glass fibres. It is a long thin bundle rather like a thin rope. At the other end of the rope is a light. The rope is called a fibre optic.

As well as using them to light things up, you can use fibre optics to look at or inside things. The light goes down through one rope, and back up into your eyes through another. In this way the insides of a hospital patient can be examined. Although quite powerful bulbs are used the light coming out of the end is quite "cold." So it is less uncomfortable for a patient in this situation.

This is a Japanese "fibrescope" used for looking inside hospital patients.

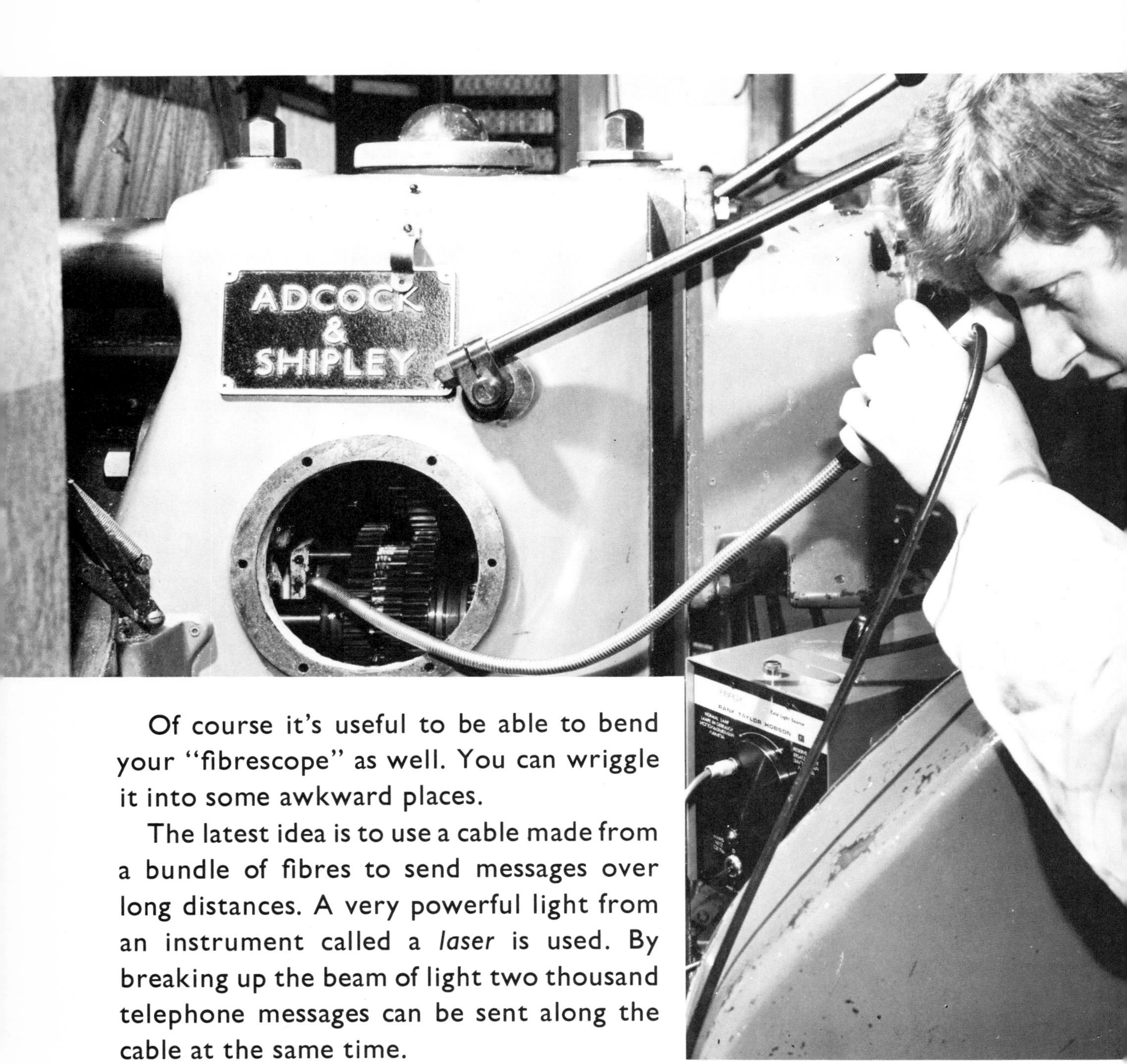

Of course it's useful to be able to bend your "fibrescope" as well. You can wriggle it into some awkward places.

The latest idea is to use a cable made from a bundle of fibres to send messages over long distances. A very powerful light from an instrument called a *laser* is used. By breaking up the beam of light two thousand telephone messages can be sent along the cable at the same time.

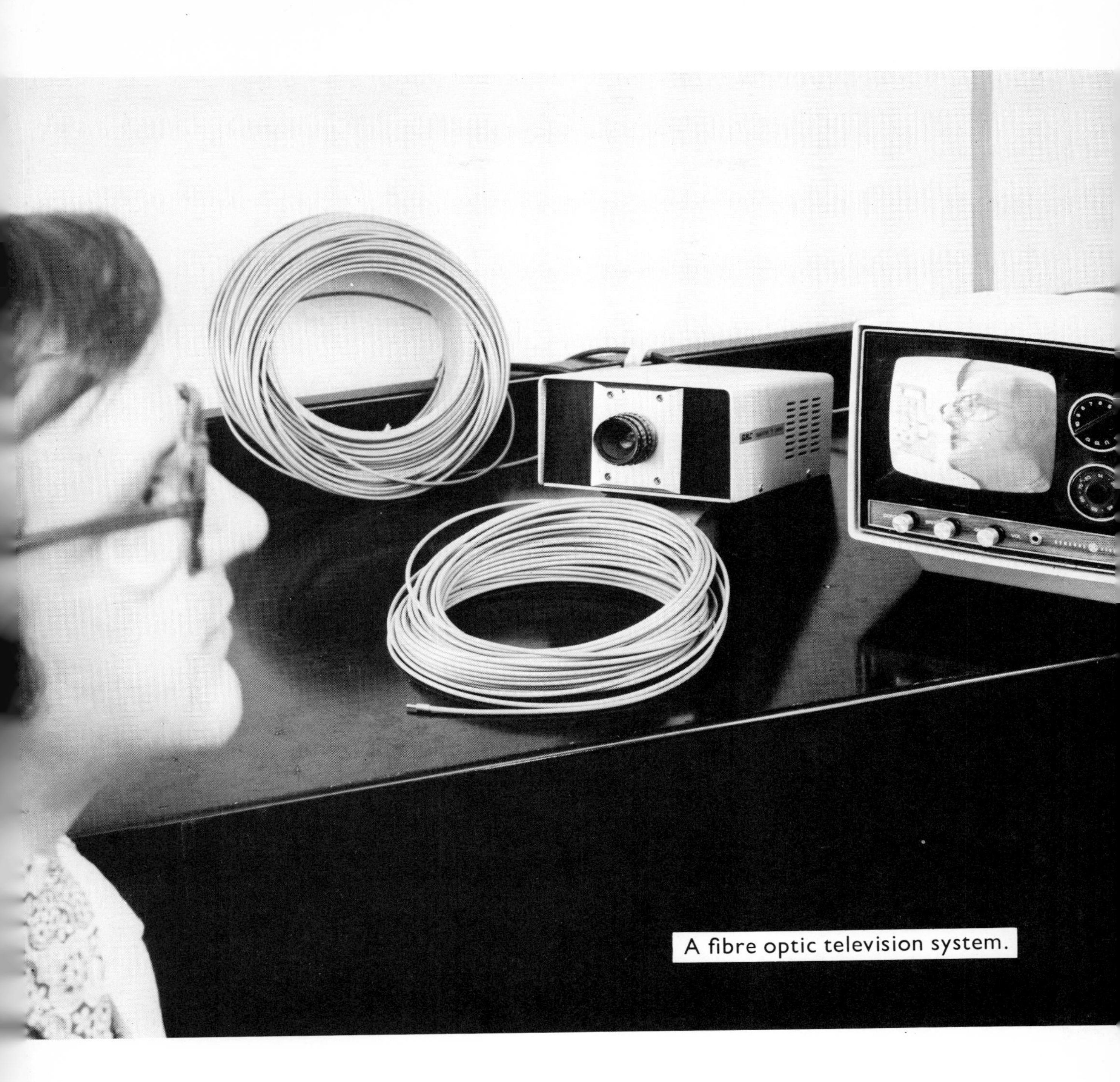

A fibre optic television system.

Very very strong and very very weak!

Above Dr. Alan Griffith (1893–1963).

In 1920 a young man was working at the Royal Aircraft Establishment at Farnborough. He was looking for the answers to two questions.

1. Glass fibres are very strong indeed when you try to break them by stretching them. How strong can they be?
2. Glass fibres will, however, break. How and why do they break?

Left A machine for testing the strength of fine fibres.

How Strong?

Using a special piece of apparatus, young Alan Griffith measured the strength of glass fibres. He tested thinner and thinner fibres and wrote down the pull it took to break each of them. He found that his best fibres were five times stronger than steel. He drew a graph like the one below.

By "stretching" his graph upwards he worked out that it should be possible to make fibres at least ten times stronger than steel. Such fibres have been made during the last few years.

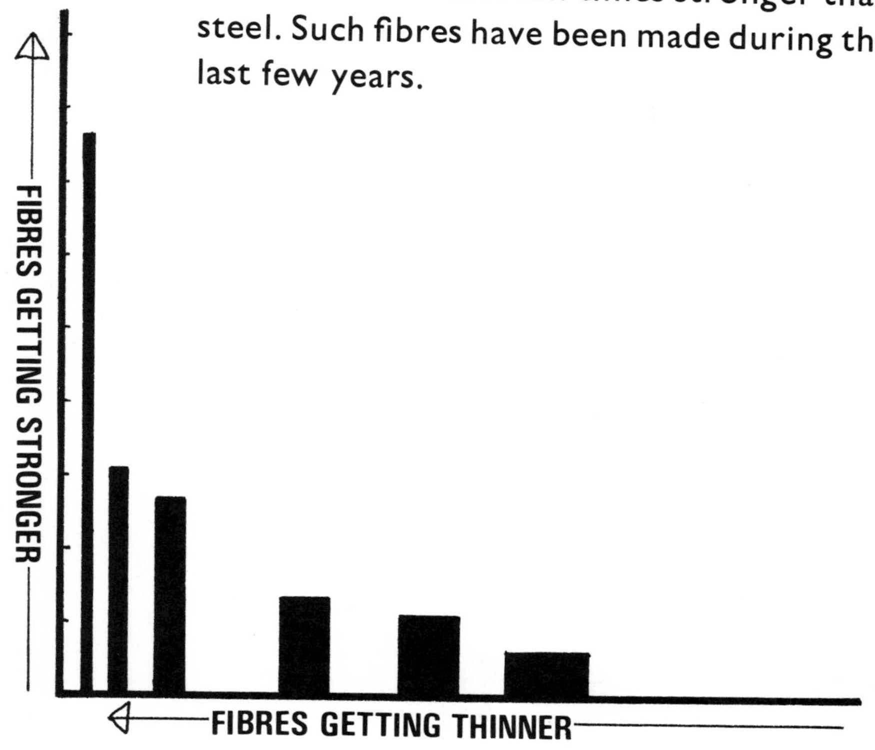

Why Are Glass Fibres Weak?

Griffith had used his experiments to work out a theory about what glass is like inside. From his theory he got the idea that glass might break easily because it was filled with lots of tiny little cracks. It's a good idea because it turns out that glass scratches and cracks much more easily than you would think. The picture shows all the cracks which have formed around a tiny scrape on a glass rod. The picture has been magnified —in real life the whole of it would fit into a full stop on this page—so the cracks are very tiny indeed.

The next part of the story happened twenty-five years later. People carrying on Griffith's work found out two important things. First, *strong fibres get weaker if you touch them.* Second, *weak fibres become stronger if you dissolve away their outsides using chemicals.*

Perhaps you can work out what this told the scientists. If not, read on . . .

Right A tiny scrape on the surface of a glass rod (Magnification x12,000).

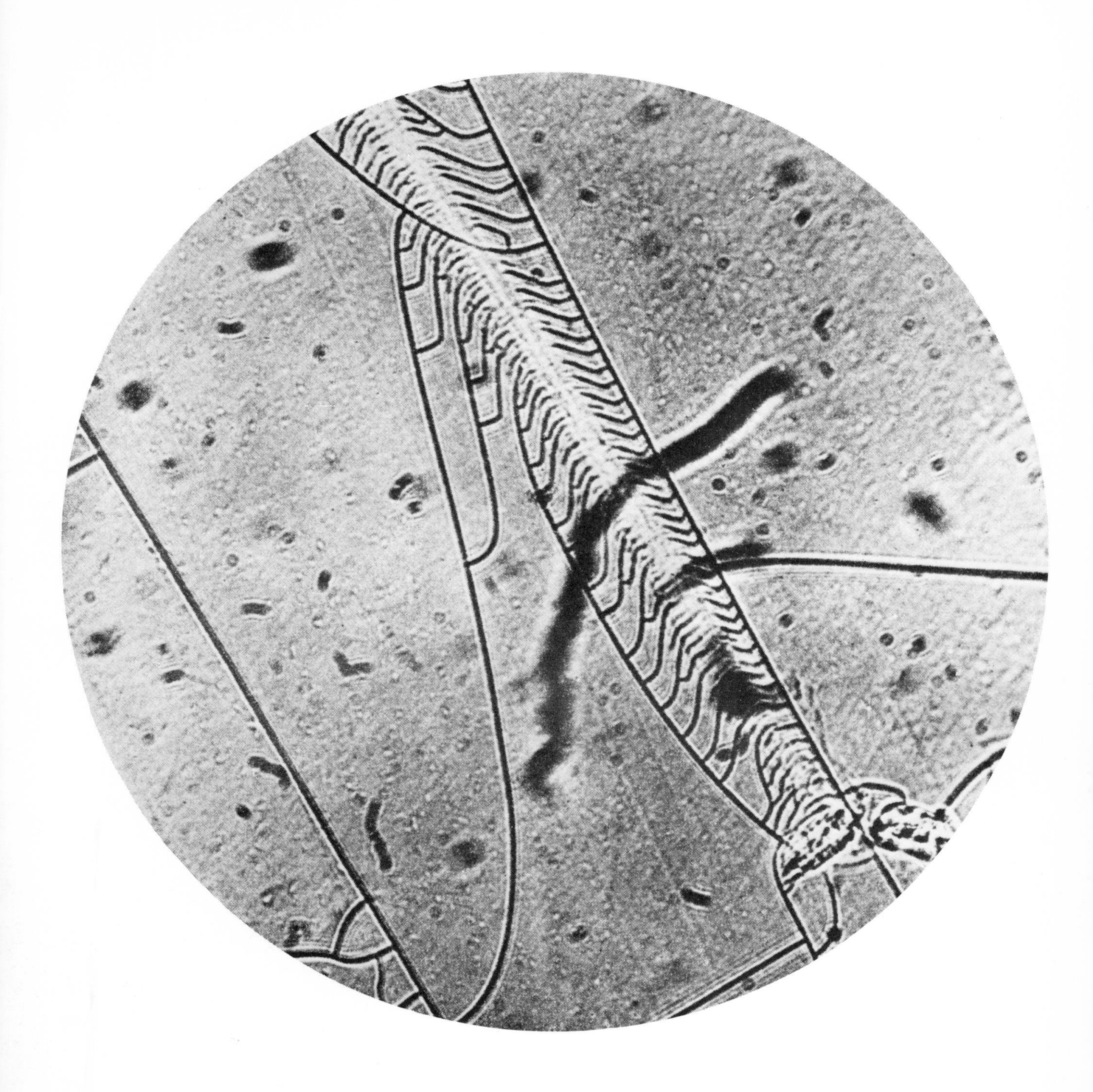

G.R.P.

The scientists working on glass fibres saw that it must be little cracks in the *outsides* of the fibres that weakened them most. Even touching a glass fibre might produce these little cracks. The answer to the problem seemed to be to cover the fibres with something that would protect them. Or—the scientists thought—why not make a new material with the glass fibres inside it, like

Below Mud huts of the Dogon people of Mali in Africa. Their buildings have remained unchanged since the twelfth century when their ancestors fled here to escape from war.

pieces of straw in mud bricks? For thousands of years people had known that putting straw in bricks made them stronger.

The substance chosen for the "mud" part of the new material was a polymer called *polyester resin*. The new material is called *glassfibre reinforced plastic*—fibreglass or G.R.P. for short. It is just one of the new *composite* materials that people have invented since. A composite material is made up from two or more other materials which stay more or less as they were when they were separate.

Perhaps these are twentieth century mud brick buildings?!

Below The "radomes" (radar domes) at Fylingdales Moor in Yorkshire, where complicated radar and other tracking equipment is protected by huge glassfibre reinforced plastic domes.

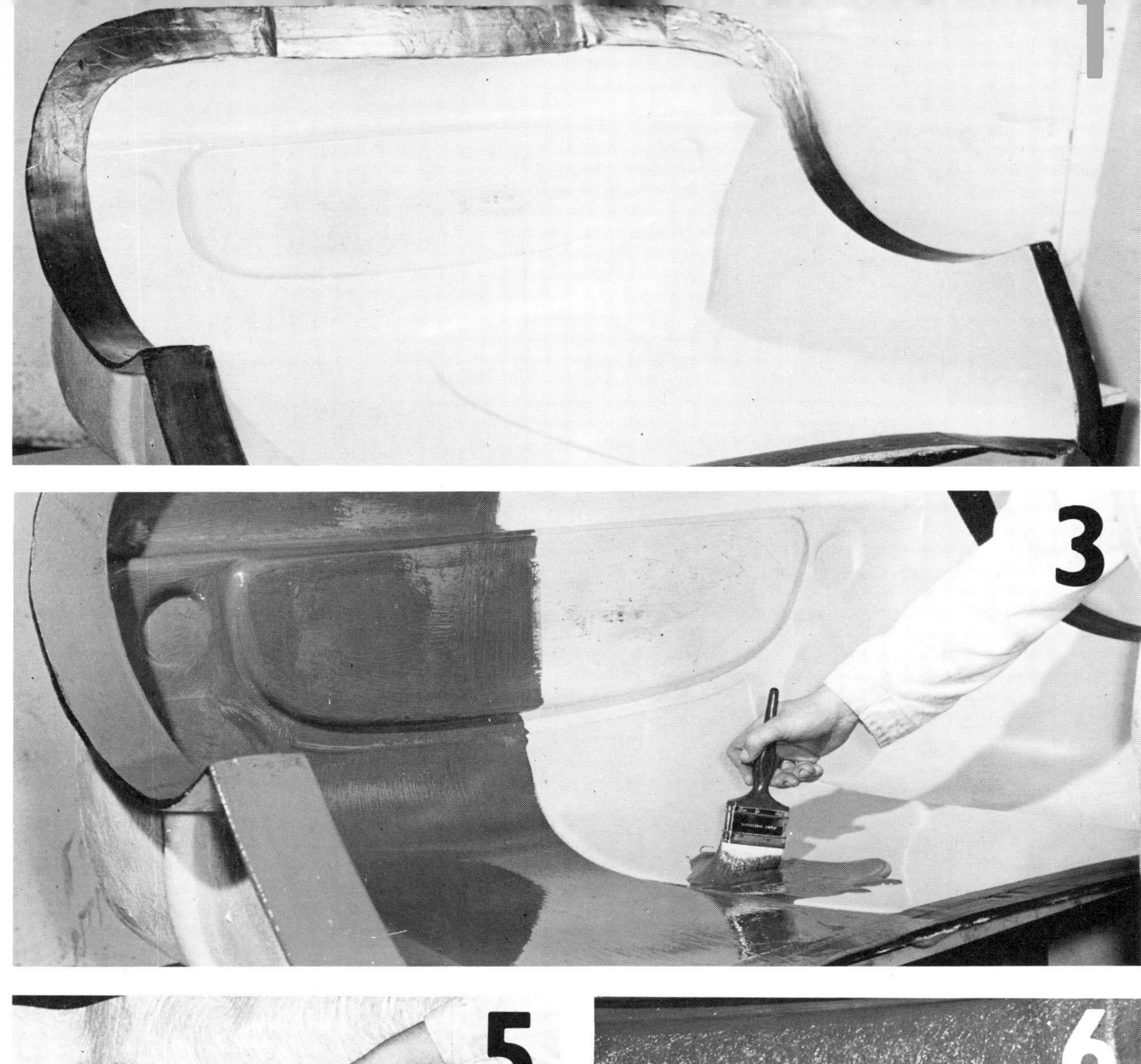
1
3

5

6

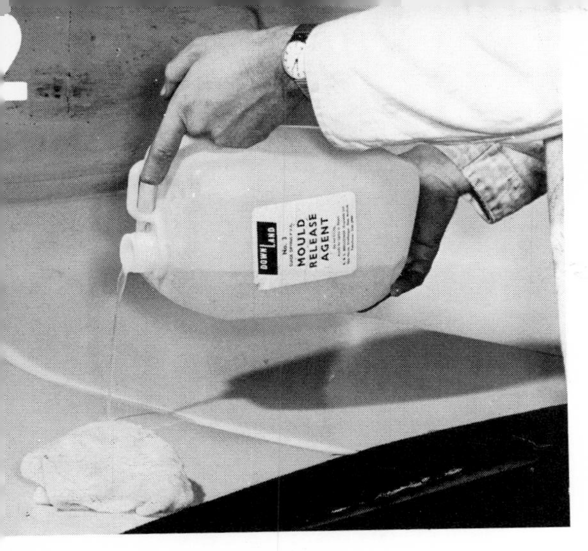

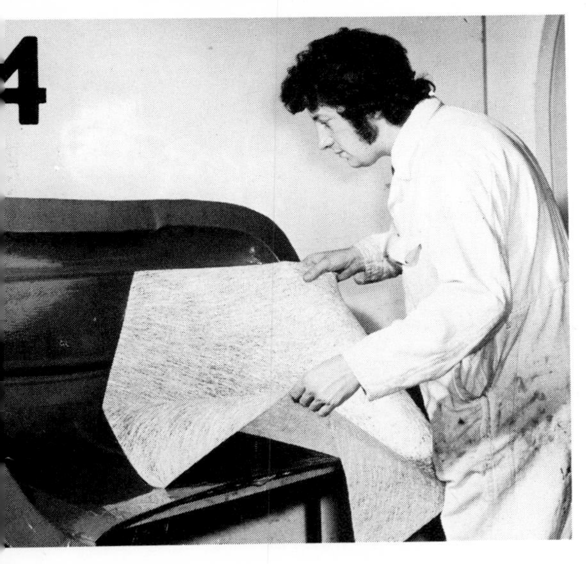

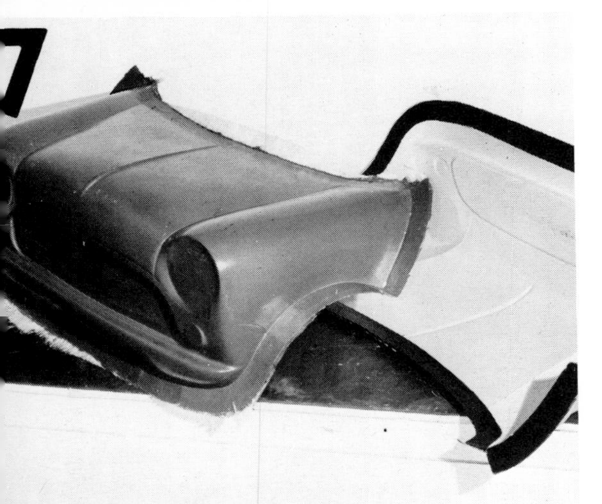

Making something out of fibreglass— it's a very simple process

1. You start with the mould. It is the shape of the moulding you are going to make but is "inside out." Usually the mould is coated with a release agent. Can you see why?
2. Now you get the resin ready. It is a matter of mixing two liquids together in measured quantities.
3 and 4. The mixture is brushed on to the mould and a piece of glassfibre mat laid into it. The piece is small enough to be a close fit.
5. More resin is "stippled" into the mat. . . .
6. This is then rolled to press out any air. (And so on, layer by layer, until the glassfibre reinforced plastic shell is thick enough.)
7. The resin begins to set. The two liquids are reacting together to form the polymer. Soon it is "cured" and can be removed from the mould.

Then all that is left is to trim and smooth the edges. Colour is usually moulded in so the moulding does not need painting, and the outside, which has been against the mould, is nice and smooth.

To do a good job, the room you work in should not be too hot, too cold or too moist. It must also be well ventilated.

H.M.S Wilton—
the only one of her kind?

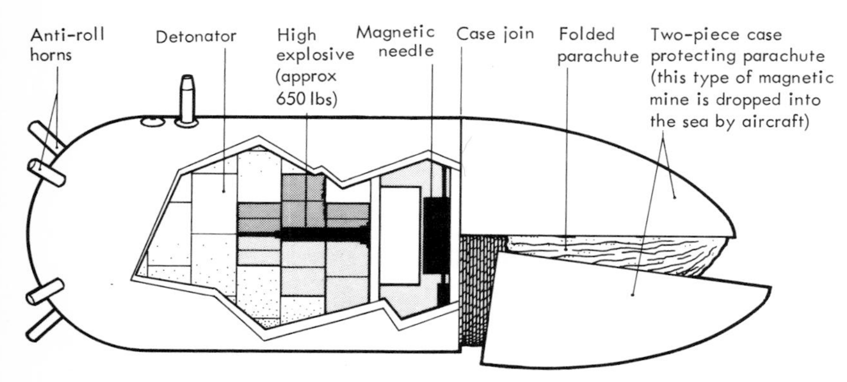

This is a magnetic mine. It is a sort of floating bomb. If anything made out of iron or steel comes anywhere near, it will explode. If you were the captain of a warship you would want to keep well away from it. That might be a nuisance. You might want to sail your ship into parts of the sea guarded by magnetic mines. And what if your ship was a mine-hunter, designed to clear away mines?

This is the Royal Navy's answer—*H.M.S. Wilton*. She is probably the largest fibreglass ship in the world. Not enough iron and steel about to set off any mines. Of course there are problems for the crew—no cans of beer, tinned food, razor blades . . .

What would we do without it?

Every day someone seems to think of a new way of using G.R.P. Here are just a few of the uses. At the bottom is a list of the properties of G.R.P. Try to decide which "use" picture goes best with each property. For example, if you think that "Very strong" is a good reason for using it to make buildings write J next to property 1. You can write more than one letter next to each property.

Properties

1. Very Strong
2. Springy
3. Light in Weight
4. Easy to make shapes from
5. Waterproof
6. Stands up to hard knocks
7. Good heat insulator
8. Built-in colour
9. Good electrical insulator
10. Good fire resistance
11. Does not rust
12. Resists attack by small animals
13. Lets light through
14. Resists attack by chemicals

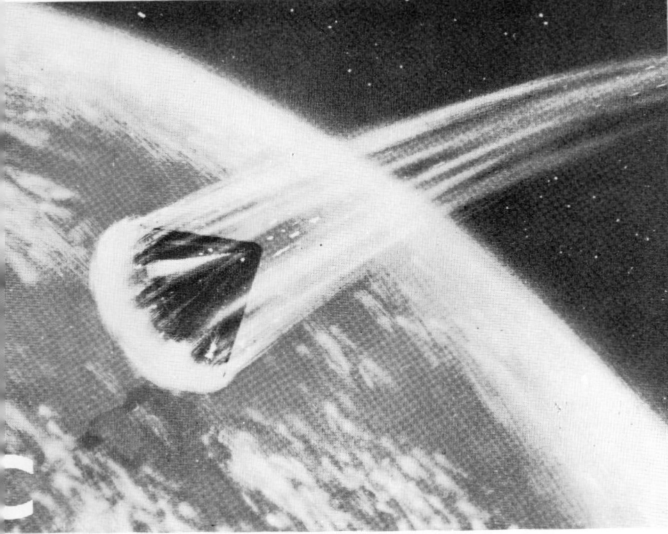

E

F

G

H

I
Medway Ports Authority

J

K

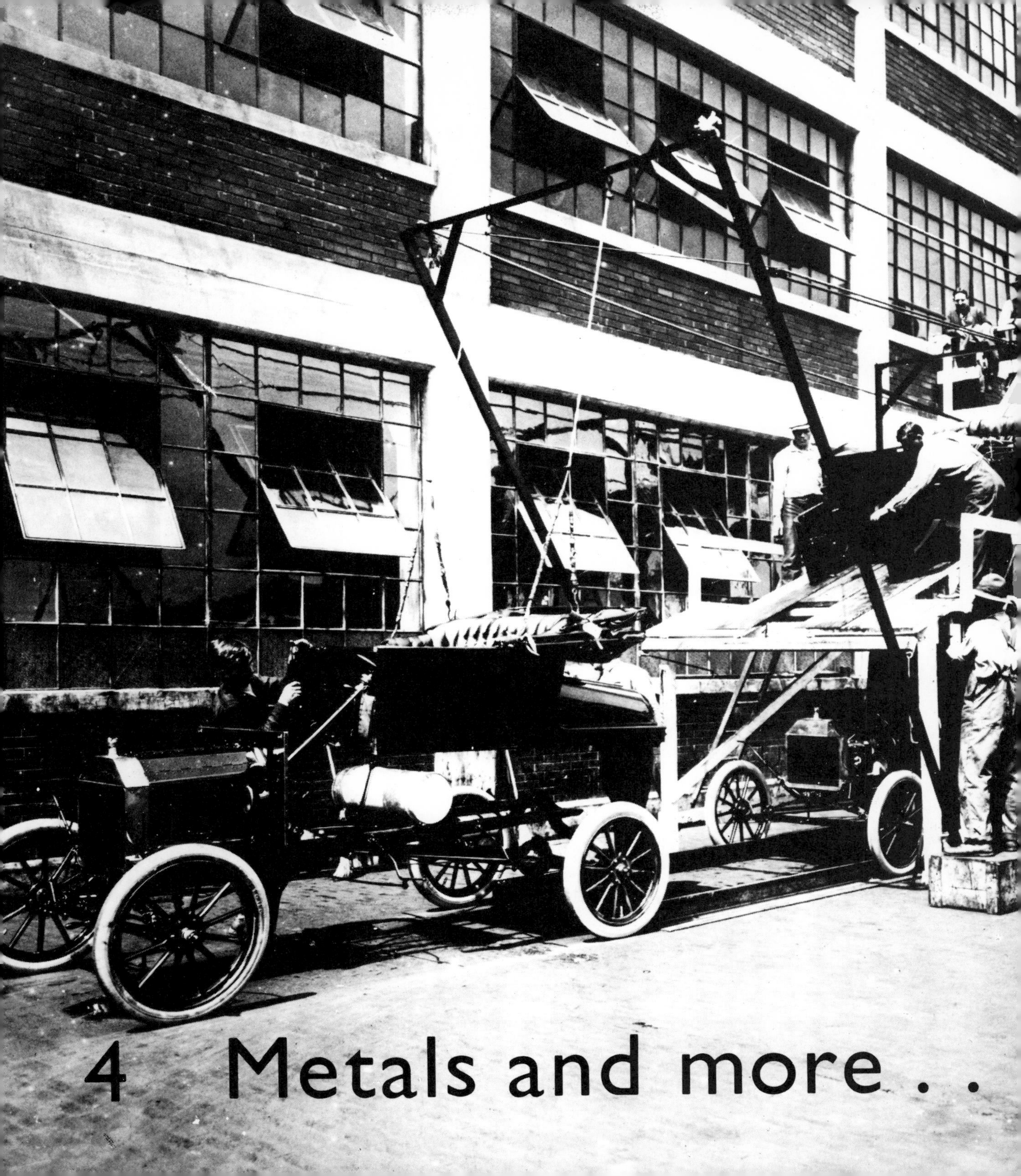

4 Metals and more . . .

What do we expect of our metals?

Strength—Chieftain army tank.

If fibreglass is so very useful—why do we go on using metals? To find the answer we can look at the things that metals can do—the ways we expect them to behave.

We expect our metals to stand up to very big pushes and pulls (forces) without tearing or breaking apart. We expect them to be *strong* and not to be *brittle*.

We expect our metals to stand up to big forces without bending too much. We expect them to be *stiff*.

We sometimes expect our metals to bend, then spring back again into their first shape. We expect them to be *elastic*.

Elasticity—the spring in this pogo stick.

Stiffness—a Dunlop tyre lever.

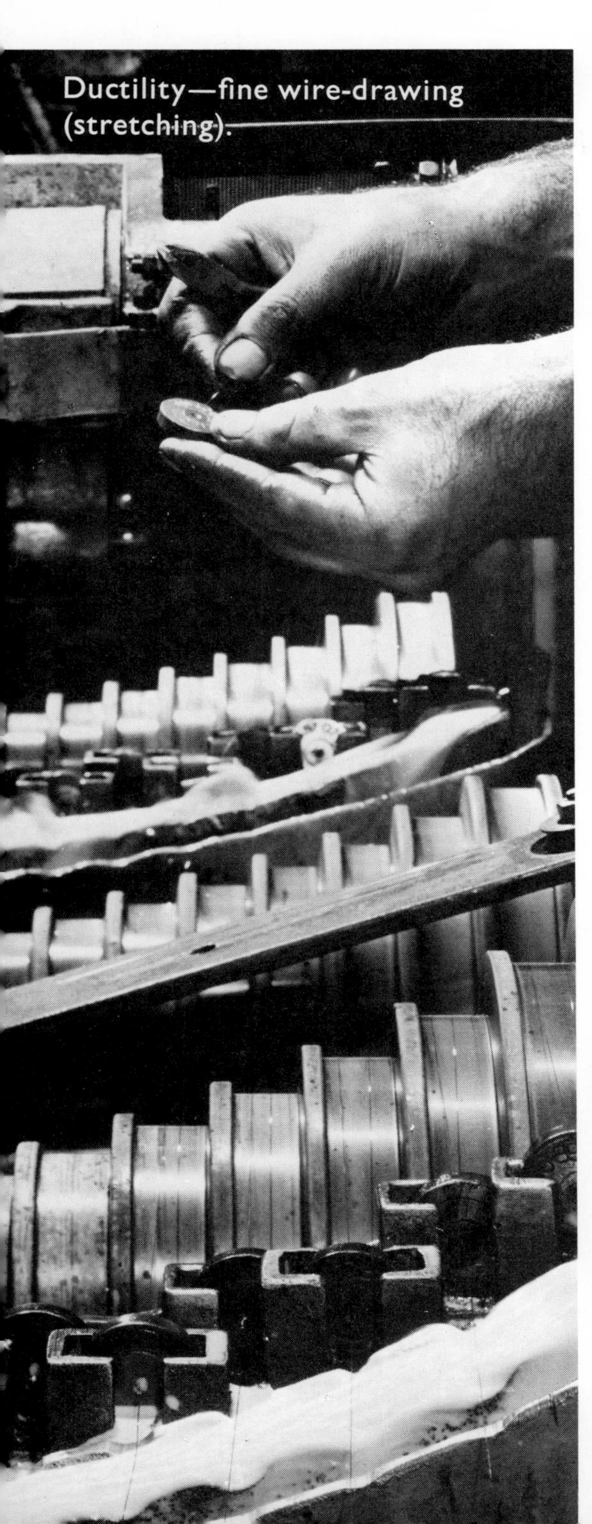
Ductility—fine wire-drawing (stretching).

At the same time . . .

We expect to be able to bang our metals into different useful shapes. We expect them to be *malleable*.

We expect that if we squeeze or pull our metals very hard, they will change into new shapes like soft toffee. Sometimes we heat them to help them do this. We expect them to be *ductile*.

Glass fibre is very strong, is ductile if heated, and is elastic. But it is not usually very stiff, and it is not malleable. So at present it can't do all the jobs that metals can.

There is another good reason why we go on using metals. Metal is easy and quick to use on the production line. Glass fibre is messy and slow.

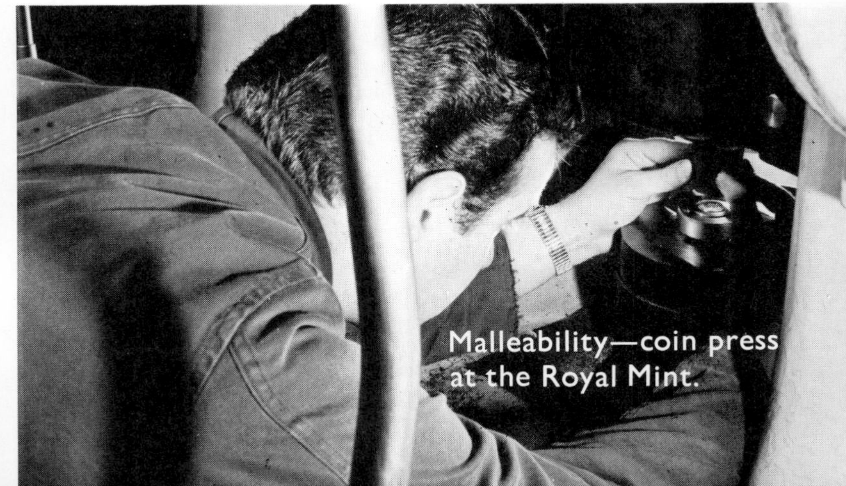
Malleability—coin press at the Royal Mint.

A theory about metals

Above Copper grains magnified x290.

Copper is the metal which most electric wires are made from. This is what it looks like under a microscope. The piece of copper is made up of lots of tiny little bits called *grains* or *crystals*.

Scientists imagine that each of the crystals is made up of the tiny little atoms we talked about on page 30.

In a "perfect" crystal the atoms are arranged like this. A crystal of iron metal would look very much like the crystal of copper but the atoms themselves would be different. A substance which is made up of atoms which are all of one sort is called an *element*. Iron and copper are elements.

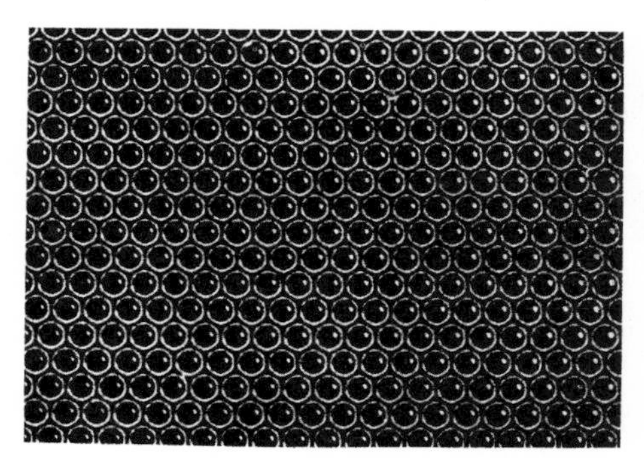

Above Bubble raft showing how atoms would be arranged in a "perfect" crystal of metal.

Scientists blow "rafts" of bubbles to study the way that the atoms might be arranged in ordinary (not "perfect") crystals of metals. They are very interested in the places where the bubbles are out of line (picture opposite). They know that in real crystals there are lots of tiny "cracks" where the atoms are out of line like the bubbles. The tiny cracks are called *dislocations*.

Above Primitive iron working.

When a piece of metal is squeezed very hard or pulled very hard the dislocations move about inside it and let it change shape. This is why metals are ductile. The trouble is that too many dislocations weaken the metal and it becomes brittle. Hundreds of years ago, and without knowing any of the theory, men worked out ways of getting just the right amount of dislocations into the metal. They also learned to treat the metals in ways that we now know made the crystals smaller. This helped to make the metals stronger.

Designing better metals

Right Iron whiskers (Magnification x20).

Today's metals scientists—called metallurgists—have used the theory to work out how strong metals ought to be. They have grown perfect crystals of metals—like these iron ones. They are "whiskers" or short fibres of iron. The whiskers are much stronger than bigger pieces of ordinary pure metal, but are very easily damaged. If the whiskers are put together in bundles they protect each other—but the bundles are only a few millimetres long. You can't make much out of little bits of metal like that!

For years metallurgists tried to think of a way round this problem. How to use the strength of metal fibres in large pieces of metal? Fibreglass provided the vital clue. Why not line up bundles of metal fibres like the glass fibres in G.R.P., and surround them with ordinary metal—like the resin in G.R.P.?

You can see the result of trying this in the picture. The material below is copper, reinforced with tungsten metal fibres. When the cracks reach a fibre they travel harmlessly along it. When the piece of metal is pulled in the direction of the fibres, it is nearly as strong as they are.

Left The black mark at the top of this microscope picture is a machined notch which would normally cause a large crack in the metal. With tungsten fibres reinforcing this piece of copper, however, a crack hardly appears.

Carbon fibres

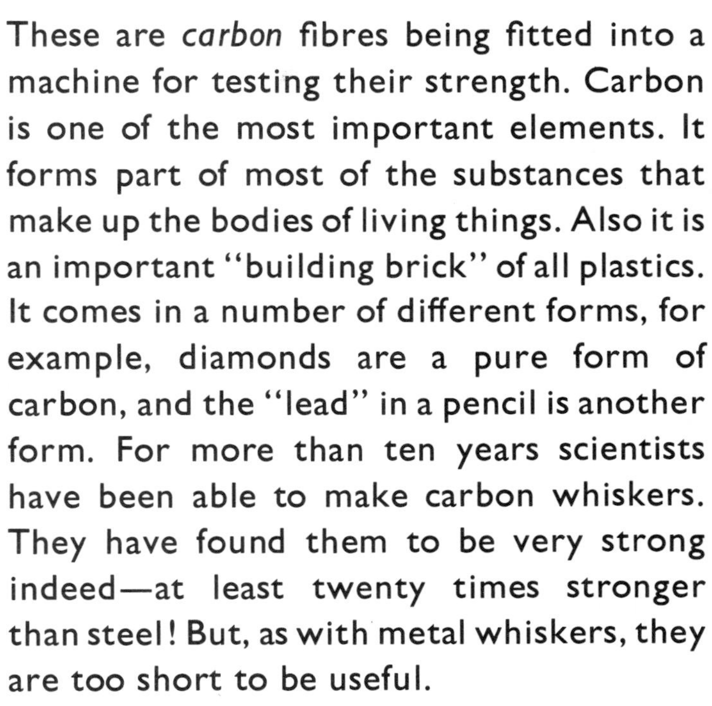

These are *carbon* fibres being fitted into a machine for testing their strength. Carbon is one of the most important elements. It forms part of most of the substances that make up the bodies of living things. Also it is an important "building brick" of all plastics. It comes in a number of different forms, for example, diamonds are a pure form of carbon, and the "lead" in a pencil is another form. For more than ten years scientists have been able to make carbon whiskers. They have found them to be very strong indeed—at least twenty times stronger than steel! But, as with metal whiskers, they are too short to be useful.

Carbon whiskers.

A carbon fibre furnace at Harwell.

In 1963 scientists at the Royal Aircraft Establishment at Farnborough began trying to work out a way of making *long* carbon fibres. They made a prototype furnace. In it, threads of a special plastic are roasted until only a thin thread of carbon is left—a long carbon fibre has been made.

On the right is a turbine—a sort of big fan which pulls air into an aircraft jet engine. The blades are made from carbon fibre reinforced plastic. This material is twice as strong as G.R.P. and many times stiffer. It is even stiffer than most metals.

Carbon fibre reinforced plastic turbine blades being fitted to a turbine.

"C.F.R.P." for strong light aircraft . . . .

. . . for golf clubs . . .

Also it is very hard wearing and light in weight. At the time the turbine was made "C.F.R.P." looked like the perfect material. In an actual jet engine it worked very well indeed until a bird was accidentally sucked into the jet intake—then the blades shattered!

Scientists are still working on the problem, and no doubt other problems will crop up. Even so, it looks as if carbon fibre reinforced materials may turn out to be some of the most important and useful ever invented. They are already being used in lots of unproblematic ways.

. . . and for fishing rods.

The oldest new material?

An important material was used to make all of these things. It was shaped whilst it was cold, then permanently hardened by baking. It was made mainly of clay or other natural earths. The material has had lots of names in its long history: "china," "pottery," "earthenware" are some of them. The modern name for the material is *ceramic*. There are lots of different kinds of ceramic made up from lots of different recipes. Some of the recipes are thousands of years old.

Today, this ancient material is being looked at by many research scientists. They believe that ceramic whiskers may be the super-fibres of the future. Used to re-inforce metals and plastics they may give us materials with unheard of properties.

Above Things made with *ceramics*: crockery, toilet bowl and cistern, spark plug. *Below* Ceramic whiskers magnified x1,000.

Copying nature—again?

Bone . . . and bamboo.

Bone is a composite material. The crystals, made of a substance called apatite, reinforce the material called collagen that they are embedded in. Bamboo is another composite. It is made from cellulose reinforced with a glassy material called silica. So perhaps all we have been doing, and will be doing, in developing new materials, is copying nature . . . from silkworms to bamboo.

Below left Bamboo magnified $\times 48$.
Below right The same piece of bamboo magnified $\times 240$.

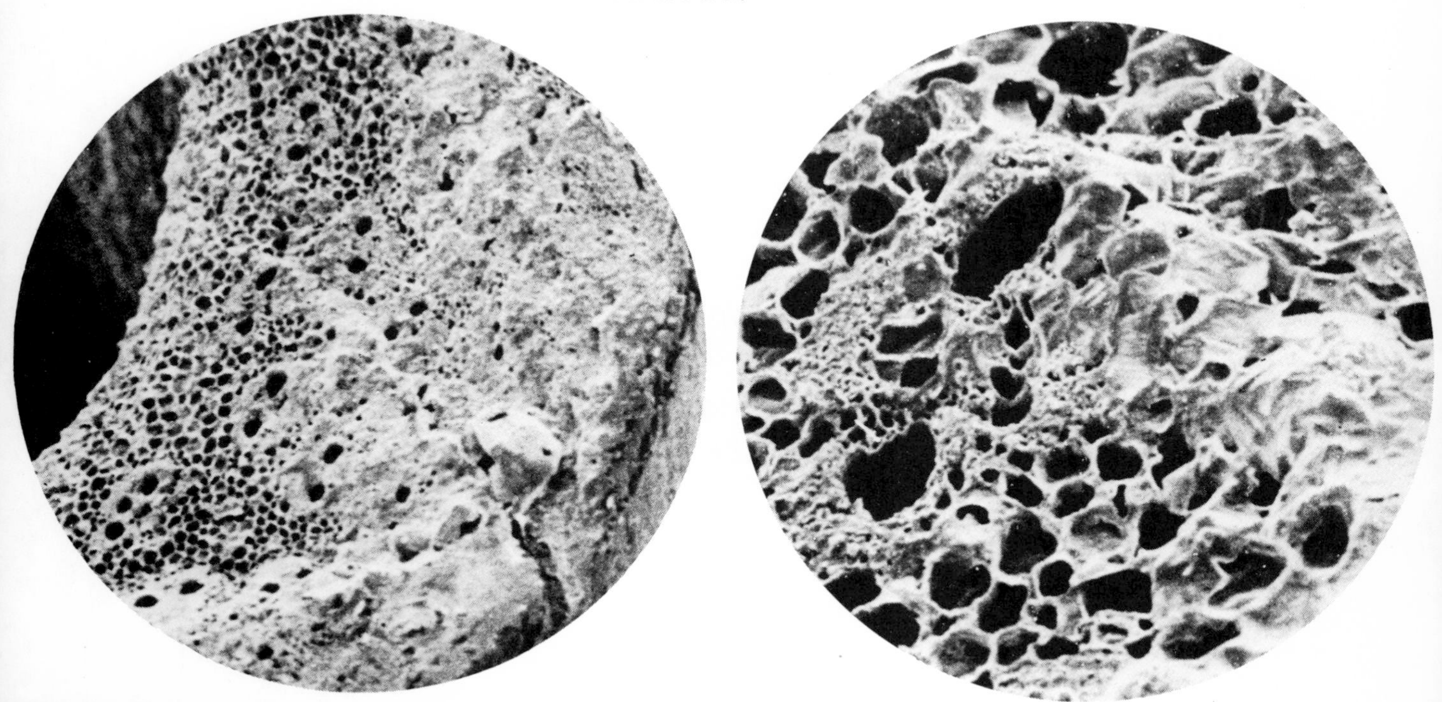

Glossary

ANIMAL FIBRES Fibres made from the coats of animals—fur and hair.

ATOMS The tiny bits of matter that scientists believe everything is made from.

BRITTLE MATERIAL A material that breaks easily and cleanly if bent or twisted.

CELLS The little "boxes" that all plants and animals are made up from.

CELLULOSE The material that the cell walls of plants are made from. It is a polymer made up of carbon and hydrogen atoms.

CERAMIC MATERIAL A material made up of mixtures of natural earths, shaped cold then baked hard.

COMPOSITE MATERIAL A material made up from two or more other materials which stay more or less as they were before being put together.

DISLOCATION A tiny, atom-sized, crack in a crystal, caused when layers of atoms slip out of line.

DUCTILE MATERIAL A material which can be pulled or squeezed into new shapes—like soft toffee.

ELASTIC MATERIAL A material which stretches or squashes when pulled or pushed—then springs back into shape again.

ELEMENT A substance made up from only one kind of atom.

FIBRE A long thin piece of material like a whisker or hair.

FIBRE OPTIC A bundle of specially treated glass fibres which act as a pipe to guide light beams along.

FILTERING Separating the solid and liquid parts of a mixture by passing it through something full of tiny holes.

G.R.P. Glassfibre reinforced (strengthened) plastic. Fibreglass.

GRAINS The tiny "building bricks" that metals are made up from. The grains themselves are made up of stacks of atoms.

KNITTING Linking up threads of material into cloth by knotting them onto each other.

MALLEABLE MATERIAL A material which can be hammered into shape without breaking.

METALLURGIST A scientist who is interested in metals.

MINERAL FIBRES Fibres made from rocks dug out of the ground.

MOLECULE A group of atoms all joined up together.

PLASTIC Short for Thermoplastic Material—a material which softens when heated then hardens again when cooled. "Thermosetting" plastics will only do this once. Plastics are usually polymers.

POLYMER A material whose atoms are joined up into long chains.

RESIN A special liquid which turns into a polymer.

STAPLE FIBRES Short fibres used for spinning into yarn.

STRONG MATERIAL A material that stands up to big forces without breaking.

TEXTILE Usually means the same as cloth.

VEGETABLE FIBRES Fibres made from various parts of plants.

VISCOSE The thick, sticky, liquid produced when natural cellulose is treated with various chemicals. Rayon is made from it.

WEAVING Linking up threads of material to form cloth by winding them in and out of each other in different directions.

YARN A long thread made by spinning fibres together.

Finding out more

Other books to read

Adler, Irving and Ruth, *Fibres* (Dennis Dobson, 1970).
Bainbridge, Jack, *Man-Made* (Evans Integrated Themes, Evans, 1973).
Farrel, Margaret, *Plastics* (E. S. A., 1962).
Harding, D. W. and Griffiths, Laurie, *Materials* (Longmans Physics Topics, Longmans, 1968).
Oliver, H. P. H., *Man-made Fibres* (Nuffield Chemistry Background Books, Longmans, 1967).
Pluckrose, H. (ed.), *The Book of Crafts* (Evans, 1971).
Shillinglaw, P., *Introducing weaving* (Batsford, 1972).
Walker, O. J., *Plastics* (Nuffield Chemistry Background Books, Longmans, 1970).
Warburton, Clifford, *The Study Book of Paper* (Bodley Head, 1967).
Wilkerson, Marjorie, *Clothes* (Batsford, 1970).

Where to write for more information

B.P. Educational Service, Brittanic House, Moor Lane, London, EC2Y 9BU.
Information Office, British Industrial Plastics Ltd., 1 Argyll St., London, W1.
Information Department, British Man-made Fibres Federation, Bridgewater House, 58, Whitworth St., Manchester 1.
Information Office, Fibreglass Ltd., St. Helens, Merseyside WA10 3TR.
Information Officer, Glass Manufacturers Federation, 19 Portland Place, London, W1.
Education Services Department, Reed Paper Group, 82 Piccadilly, London, W1.
Shell Education Service, Shell Centre, London, SE1.
United States Information Services, PO Box 2LH, Upper Brook St., London, W1. (*Insight* magazine.)

Books for teachers

Chubb, L. W., *Plastics, rubbers and fibres: materials for man's use* (Pan, 1967).
Couzens, E. G. and Yarsley, V. E., *Plastics in the Modern World* (Penguin, 1969).
Fichlock, D., *New Materials* (Murray, 1967).
Gordon, J. E., *The New Science of Strong Materials* (Penguin, 1968).
Green, D., *Experimenting with Pottery* (Faber, 1971).
Pilkington, R., *Glass* (Signpost Library Chatto, 1965).
Scientific American, *Materials* (Freeman, 1967).

Index

Acknowledgements

The author and publishers thank the following for their kind permission to reproduce the pictures that appear on the following pages. B.I.C.C. Ltd., 73 left; Birmingham University, 83; Bristol Uniforms, 8 bottom left; British Ceramic Research Association, 82 bottom; British Hovercraft Corporation, 47 top; British Petroleum, 15 bottom left; Cambridge Scientific Instruments Ltd., 10, 11, 79 right; Cape Fibres, 16; Central Office of Information, Crown copyright, 72 top, 73 right; Copper Development Association, 74 top; Corning Glassworks, 54, 57; Courtaulds Ltd., *frontispiece*, 6, 7, 8 bottom left, 17, 22, 23, 24, 25, 26, 27, 28, 38, 39, 40 left, 44, 45 left, 80 right, 81 top and left; Doulton Sanitaryware Ltd., 82 top centre; Down Brothers, 46 top; Du Pont Ltd., 32, 35 top left, 41 left; Dunlop Ltd., 72 bottom right; F.A.O., 14 centre left; Fibreglass Ltd., 48, 50 top, 51 top, 51 bottom right, 52, 53, 64, 65, 69 top and bottom left and bottom right, 70 top left, 70 bottom centre and right; Ford Ltd., 71, 72 bottom right; Fothergill & Harvey Ltd., 50 bottom right, 69 right top, 70 bottom left; Fuji Photo Optical Co. Ltd., 55 bottom; Ken Goodall, 68; I.C.I. Fibres Ltd., 8 top, 9 top right and bottom right, 34, 35 top centre, top right and bottom, 40 right, 43; Irish Linen Guild, 14 centre right; Mary Evans, 12 bottom, 76; Mary Quant, 45 right; Ministry of Defence, Crown copyright, 67; Morganite Modmor Ltd., 81 right; NASA, 69 left centre; National Physical Laboratory, Crown copyright, 78; Owens-Corning Ltd., 51 bottom left; Pasteur Institute, 18, 21; Pictorial Press, 15 top, 15 centre left, 15 centre right, 72 bottom left; Pilkington Brothers Ltd., 49; Radio Times Hulton Picture Library, 33; Rank Optics, 50 left, 55 top, 56; Reliant Ltd., 70 top right; Rolls Royce Ltd., 8 bottom right, 58 top; Royal Society, 61; Science Museum, 13 bottom; Shell Ltd., 42, 46 bottom, 47 bottom; Shirley Institute, 15 bottom right, 19 bottom, 36, 37; Smiths Industries Ltd., 82 top right; T.I. Research Labs., 58 bottom; Transcontinental Films, 9 top left; Tropical Products Institute, 14 top; U.K.A.E.A., 79 left, 8 left; Union Carbide, 31; University Museum of Arch. & Ethn. (Cambridge), 12 top; Wedgwood, 82 top left; Westinghouse Ltd., 75; W.H.O., 62; *Yorkshire Post*, 63.